ALGEBRAIC TOPOLOGY

WOLFGANG FRANZ

Professor of Mathematics

University of Frankfurt-on-Main

Translated from the German by

LEO F. BORON

University of Idaho

With the collaboration of

SAMUEL D. SHORE

University of New Hampshire

JAMES J. ANDREWS

HARRY F. JOINER II

ROBERT C. MOORE

The Florida State University

KYOSHI ISÉKI

Kobe University

FREDERICK UNGAR PUBLISHING CO.

NEW YORK

Translated from the German original
Topologie II: Algebraische Topologie

By arrangement with
Walter de Gruyter & Co., Berlin

Printed in the United States of America

Library of Congress Catalog Card No. 65-28047

CONTENTS

Part II

CHAIN COMPLEXES AND THEIR APPLICATIONS

Part III

CELL COMPLEXES. INVARIANCE

Part IV

DEVELOPMENT OF THE THEORY

INTRODUCTION

§ 1. Analytic and Algebraic Topology

In Volume I, *General Topology*, we discussed the most general possible classes of spaces and derived their properties deductively from a small number of axioms. One may also therefore designate this branch of topology as "Analytic Topology." Accordingly, the assertions of general topology are very extensive and general, and for that reason relatively non-concrete. In the present Volume II we treat the other branch of contemporary topology, *Algebraic Topology*. Here we will deal with more special problems and shall place particular emphasis on concepts and theorems which are closely related to direct geometric intuition. In this connection, we must essentially limit the domain of the basic spaces within our framework: We consider polyhedra almost exclusively.

The real difference between the two branches of topology consists in the method. The nomenclature "Algebraic Topology" indicates that algebraic tools are drawn upon for the treatment of topological questions. We can explain even at this point in rough outline the procedure to be used for this. We consider a so-called *category* $\mathfrak{R}$ which consists of spaces and continuous mappings or, more specifically, of the spaces R, S, . . . of a fixed class (for instance, of all Hausdorff spaces or of all compacta or as in our case of all polyhedra R, S, . . .) and of all continuous, not necessarily monomorphic or epimorphic mappings $f: R \to S$ among them. By means of a procedure which will be developed later, we shall assign to each space R of $\mathfrak{R}$, for each natural number $q = 0, 1, \ldots$, an abelian group $H_q(R)$, the so-called q-dimensional homology group of R. Furthermore, for each q, we shall assign to each continuous mapping $f: R \to S$ a homomorphism $f_* : H_q(R) \to H_q(S)$, the f-induced homomorphism f_* of the corresponding homology groups. This double correspondence—groups to spaces and homomorphisms to mappings—has the following properties: (I) If $i: R \to R$ is the identity mapping, then, for each q, i_* is the identity homomorphism of $H_q(R)$. (II) If $f: R \to S$ and $g: S \to T$ are continuous mappings of spaces in $\mathfrak{R}$ and $g \circ f: R \to T$ is the composition mapping, then $(g \circ f)_* = g_* \circ f_*$; the composition mapping thus corresponds to the composition homomorphism. One thinks of the class of all abelian groups H, G, . . . together with all their homomorphisms $\varphi: H \to G$ also as a *category* $\mathfrak{H}$. A mapping of $\mathfrak{R}$ to $\mathfrak{H}$ which assigns to each space $R \in \mathfrak{R}$ a group $H \in \mathfrak{H}$ and to each mapping $f \in \mathfrak{R}$ a homomorphism $f_* \in \mathfrak{H}$ so that the properties (I) and (II) are satisfied is called a *functor from* $\mathfrak{R}$ *to* $\mathfrak{H}$. The correspondence we described is of this nature; it is called the *homology functor* H of $\mathfrak{R}$ to $\mathfrak{H}$. The homology functor H translates topological questions about spaces and mappings of the category $\mathfrak{R}$ into algebraic questions about groups and homomorphisms of the category $\mathfrak{H}$. The definition and study of this homology functor H and its generalizations and refinements form the content of homology theory, which in turn is the most important basic idea of algebraic topology. The study of H within the framework given here also forms the content of this volume.

In a certain way, the method of algebraic topology imitates that of analytic geometry, in which geometric facts are also translated into algebraic facts. However, an essential difference is that the mapping by means of the homology functor H is not one to one. Different spaces R do not necessarily have different homology groups $H_q(R)$; different continuous mappings f do not necessarily correspond to different f-induced homomorphisms f_*. Since not all topological objects can be represented in this way by means of corresponding algebraic objects, we can look at this as a disadvantage. On the other hand, it is also an advantage insofar as we can translate in this way difficult and perhaps seemingly unsolvable geometric problems into simpler, but perhaps in compensation solvable, algebraic questions and we can thus arrive in any case at partial results. In particular, we can hope thus to arrive at criteria which can be established or even can be used for computation purposes. In any case, we observe here that in the investigation of topological questions the functor H leads generally only to necessary but not always to sufficient algebraic criteria. It depends on the geometric problem at hand whether one arrives by means of a stepwise refinement of the algebraic methods finally to a complete solution and to sufficiency criteria. Recently, several remarkable results in this direction have been attained. In the development of algebraic tools for topology, methods have evolved which are of independent algebraic interest and which have found a realization in "Homological Algebra." But we point out, as we also did in the Introduction to Volume I, that we are concerned here exclusively with the attainment of geometric-topological knowledge.

If we assign algebraic objects to topological spaces or, more generally, to arbitrary topological structures, as happens for example by means of the homology functor H, then the question of the *invariance* of this object always appears. We clarify this question by the example of the homology group $H_q = H_q(R)$, which we assign to a polyhedron R. (The construction of $H_q(R)$ will proceed from a special representation of the space R.) But the construction must be made so that if we take as starting point another space R' which is homeomorphic to R and apply the same method of construction to R' then the same or, in any case, an isomorphic homology group results, i.e., $H_q \approx H_q(R')$. It is only in this way that we arrive at the fact that H_q is assigned in an invariant way to the space type of R, that is, it is independent of the special representation of R, or, as we say, $H_q(R)$ is an *invariant* of the space type of R. Already in general topology we have become acquainted with invariant properties, for example, the concepts of being Hausdorff, regular, normal, and compact. Also we have recognized the dimension dim R of a compactum R in Volume I, §32, expressly as an invariant of the space type R.

In the theory of polyhedra, the question of invariance plays a special role. A polyhedron R is determined with the aid of a simplicial complex K, that is, as the body $R = |\mathsf{K}|$ of the latter. The building blocks of K, the simplexes, can be seen without more explanation. All topological properties of the space must therefore be reflected in the manner in which these finitely many building blocks are put together, in other words in the incidence relations among these simplexes. Since we are dealing here with only finitely many incidences, there thus results a certain combinatorial problem. *Combinatorial topology* sets for itself the charming as well as difficult problem of deriving from the finite combinatorial schema of the incidences of K the properties of the polyhedron $R = |\mathsf{K}|$. The main difficulty centers in the invariance question. That is, if K' is another simplicial decomposition of the polyhedron $R = |\mathsf{K}| = |\mathsf{K}'|$, then the mutual disposition of the two simplicial complexes K and K' which we pointed out already in Volume I, §30, is in general so complicated that a deduction from one applicable to the other is very difficult. The so-called "Principal conjecture of combinatorial topology" indeed asserted that there are subdivisions K_1 of K and K_1' of K' such that K_1 and K_1' are isomorphic. But this conjecture, after remaining open for many decades, appears not to be valid. In any case, it cannot be made the basis of combinatorial topology. One is therefore prompted to search for properties of K which are invariants of the polyhedron $R = |\mathsf{K}|$, that is, properties which can be read off in the same way from every other simplicial decomposition K' of R. The question whether or not a given property of K is invariant and perhaps the proof of invariance will also play an important role in our development and will at times require strenuous effort.

We can illuminate the concept of topological invariance from still another, somewhat more special viewpoint. For instance, let the topological space R be a subspace of a Euclidean n-space R^n. Let R be carried over, by means of an isotopic deformation inside of R^n, into the space R', that is, by means of a deformation in which the image of R also remains homeomorphic to R for all intermediate positions. The topological invariants of R are naturally also invariants of R', they appear now as invariants with respect to isotopic deformations. One might therefore be tempted to characterize topology as the theory of invariants with respect to isotopic deformations, to think of it in a certain way as the geometry of elastic deformations or as "rubber geometry." This conception is, above all, according to what was asserted before, somewhat too narrow and not entirely suitable due to its coarseness. But it yields nonetheless an intuitive picture required for the first approximation of the tendency of topological investigations.

In this volume only a small outline of algebraic topology can be given. We shall deal chiefly with the foundations of *homology theory* which represents the oldest, most extensively developed branch of algebraic topology and today is indispensable for many parts of mathematics. In the following §2 of this introduction we introduce a number of important topological problems and explain them by examples to which we shall return later from the standpoint of homology theory.

§ 2. Problems and Examples

We consider mappings $f: R \to S$ of a topological space R into a topological space S, which without its always being asserted explicitly are assumed to be continuous. The topological spaces R and S are later taken to be polyhedra. The mappings f need be neither monomorphic nor epimorphic (Volume I, §5). Among them, the *constant mappings* play a special role; they map every point of R into the same point $q \in S$. The mappings $f: R \to R$ are called *mappings of R into itself*. Among them, the *identity mapping* $i = i_R$, which maps each point of R into itself, plays a special role. If $A \subset R$ is a subspace of R, then the mapping $f: A \to R$, which maps each point of A into itself, is called the *inclusion mapping* of A into R; it must of course be distinguished from $i_A: A \to A$.—We first point out the two most important problems for topological spaces, the homeomorphism problem and the homotopy problem.

A. The homeomorphism problem. Two spaces R and S are called homeomorphic, in symbols $R \approx S$ (cf. Volume I, Def. 5.1), if there exists a one-to-one bicontinuous function from R onto S which carries over the topological structure of R (for instance, the open sets or the neighborhoods in R) into the topological structure of S. The theorem (Volume I, Theorem 5.6) that R and S are homeomorphic if and only if there exists a one-to-one, that is, monomorphic and epimorphic, mapping $f: R \to S$ which is continuous as well as its inverse $f^{-1}: S \to R$ is valid. If R is a compactum, in particular a polyhedron, then one can omit the last-named requirement on f^{-1}, since in this case it follows from the other requirements (Volume I, Theorem 18.2).

The homeomorphism problem asks whether or not two given topological spaces R and S are homeomorphic. If we assign the same space type to two homeomorphic spaces, it is therefore to be decided whether R and S belong to the same space type or not. It is further required to produce, for each space type, a special space R as representative, also called the normal form. For the time being, the homeomorphism problem is unsolved in this form. In the sequel, we shall be able to produce with the aid of homology theory only relatively coarse necessary conditions for the homeomorphism of two spaces. We shall give a solution of the homeomorphism problem including the construction of normal forms only for the very special class of two-dimensional surfaces.

B. The homotopy problem

2.1 Definition: Two continuous mappings $f, g: R \to S$ are called *homotopic*, denoted $f \simeq g$ or more precisely $f \simeq g: R \to S$, if either one of the following two equivalent conditions is satisfied:

(1) There exists a family of mappings $F(p, t): R \to S$, which are continuous in the two variables $p \in R$ and the real number t in the interval $0 \leq t \leq 1$, such that $F(p, 0) = f(p)$ and $F(p, 1) = g(p)$.

(2) There exists a continuous mapping $F: R \times I \to S$ of the

product space $R \times I$ of R and the unit interval $I \equiv \{t \mid 0 \leqq t \leqq 1\}$ into the space S such that for each point $p \in R$, $F(p, 0) = f(p)$, $F(p, 1) = g(p)$ hold.

The family of mappings $F(p, t)$ or the mapping F is called a *homotopy*, more specifically, a *homotopy connecting f and g*.

We must note than in (1) the simultaneous continuity in both the variables p and t is required.—That (1) and (2) are equivalent is obvious.—We also say briefly that f is continuously deformable into g. The adjacent figure shows a sketch for the case when R is a segment and S is a rectangle; the dotted curves point out the paths which the image $F(p, t)$ of a fixed point $p \in R$ describes if we allow t, interpreted, for instance, as time, to range from 0 to 1.

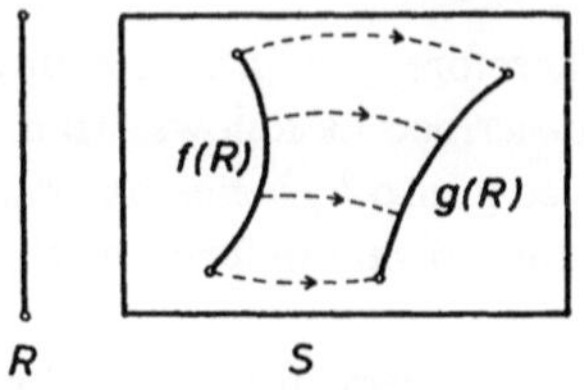

Fig. 1

Examples: (a) Let both R and S be the Euclidean plane R^2, $f = i_R$ the identity mapping, and g be such that $g(x, y) = (0, 0)$, the constant mapping into the point $(0, 0)$. We have $f \simeq g$; a connecting homotopy is $F((x, y), t) = ((1 - t)x, (1 - t)y)$.—(b) Let both R and S be the n-sphere S^n $(n \geqq 1)$, $f = i_R$ the identity mapping, g the antipode mapping which carries each point in S^n, interpreted as a unit sphere in Euclidean space R^{n+1}, into its opposite (i.e., antipodal) point. For $n = 1$, we obviously have $f \simeq g$.—

Counterexamples: (c) Let $R = S$ and let f and g be defined as in (b). For $n = 2$, we do not have $f \simeq g$, as we shall see in §23.—More generally, in (b) we have $f \simeq g$ for odd n, $f \not\simeq g$ for even n.—(d) For the same given data as in (b), let $f = i_R$ and let g be a constant mapping. Then $f \not\simeq g$.

2.2 Theorem: *The relation of homotopy among the mappings of a space R into a space S is reflexive, symmetric and transitive.*

Proof: $F(p, t) = f(p)$ for all points $p \in R$ and all $0 \leqq t \leqq 1$ is a self-connecting homotopy of $f\colon R \to S$; we therefore have $f \simeq f$.— If $F(p, t)$ is a homotopy connecting $f\colon R \to S$ with $g\colon R \to S$, then $F'(p, t) = F(p, 1 - t)$ is a homotopy connecting g with f; it therefore follows from $f \simeq g$ that $g \simeq f$.—Now let $F(p, t)$ be as just defined and let $G(p, t)$ be a homotopy connecting g with $h\colon R \to S$. Then, as one can easily verify, the homotopy

$$H(p, t) = \begin{cases} F(p, 2t) \text{ for } 0 \leqq t \leqq \frac{1}{2}, \\ G(p, 2t - 1) \text{ for } \frac{1}{2} \leqq t \leqq 1 \end{cases}$$

connects f with h. Therefore, $f \simeq g$, $g \simeq h$ imply that $f \simeq h$.—The proof can be briefly described as follows: In a unit of time, one can deform f into g and also g into h; hence one can also deform f into h in the same unit of time, if one deforms with doubled speed first f into g and then g into h.

On the basis of this theorem, for given R and S, the totality of mappings $f\colon R \to S$ decompose into classes of homotopic mappings, the so-called *homotopy classes* or *mapping classes*. The homotopy problem asks, for given R and S, whether two mappings $f, g\colon R \to S$ are homotopic and furthermore to enumerate the totality of mapping classes of R and S, for instance by giving the normal forms. In this generality, the problem is also unsolved for polyhedra R and S. It is only in very special cases, for example for spheres $R = S = S^n$, that one knows all mapping classes. We will become acquainted with examples in §23.—We prove further the following theorem which will be important later.

2.3 Theorem: If $f \simeq f'\colon R \to S$ and $g \simeq g'\colon S \to T$, then $g \circ f \simeq g' \circ f'\colon R \to T$.

Proof: If $F(p, t)$ is a homotopy connecting f with f', then $g(F(p, t))$ is obviously a homotopy connecting $g \circ f$ with $g \circ f'$. We therefore have that $g\circ f \simeq g\circ f'$. Analogously, we have $g\circ f' \simeq g'\circ f'$, from which the assertion follows according to the preceding theorem.

C. Special homotopies. We select from among the numerous applications of the homology concept several which are particularly simple and indicate the corresponding problems.

1. *Inessential Mappings*

2.4 Definition: A mapping $f\colon R \to S$ is called *inessential* if it is homotopic to a constant mapping $c\colon R \to S$. Otherwise, it is called *essential.*

Examples: (a) A mapping $f\colon R \to S^n$ of a space R into the n-sphere which is not epimorphic is inessential. That is, if $q \in S^n$ is not in the image of f and q' is the antipode point of q, then f is homotopic to c the constant mapping into the point q'. Namely, if $p \in R$, then the points $f(p)$ and q' are not antipode points; they therefore uniquely determine a greater arc of a great circle and a smaller arc of a great circle on S^n; if one lets $f(p)$ run uniformly with t over this arc from $f(p)$ to q', then one obviously obtains a homotopy connecting f with c.—(b) The identity mapping of the n-sphere is essential, as we shall show later.

2. *Deformations of the Identity*

2.5 Definition: A continuous mapping $f\colon R \to R$ of R into itself with $f \simeq i_R$ is called a *deformation of the identity.*

The antipode mappings of the n-sphere form, for odd n, an example of deformations of the identity, but not an example when n is even.

3. *Contractible Spaces*

2.6 Definition: A space R whose identity mapping i_R is inessential, that is, is homotopic to the constant mapping into a point $p \in R$, is called *contractible*, more specifically *contractible into p.*

Euclidean n-space R^n as well as a simplex $[\sigma]$ is contractible. The n-sphere is not contractible.

4. *Retracts*

2.7 Definition: A subspace $A \subset R$, for which there exists a continuous mapping $r: R \to A$ with $r(a) = a$ for all $a \in A$, is called a *retract* of R. The mapping r is called a *retraction* or a *retracting mapping*.

If $i: A \to R$ denotes the inclusion mapping, then $r \circ i = i_A$, the identity mapping of A.—Examples: (a) A single point $p \in R$ always represents a retract of R; the retracting mapping is the constant mapping into p.—(b) The product space $R \times S$ of two spaces R and S has R and S as retracts. The retracting mappings are the two projections of the product space onto R and S.—A *counterexample* is a disk R of the Euclidean plane and its one-dimensional boundary sphere $A = S^1$; S^1 is not a retract of R, as we shall see in §23.

5. *Deformation Retracts*

2.8 Definition: A retract $A \subset R$ with the retracting mapping $r: R \to A$ and the inclusion mapping $i: A \to R$ is called a *deformation retract* of R if the mapping $r_0 = i \circ r$ of R into itself is homotopic with the identity mapping i_R.

Therefore, for a deformation retract, the retracting mapping r, more specifically the mapping $i \circ r$, is a deformation of the identity. Thus, the equations $r \circ i = i_A$ and $i \circ r \simeq i_R$ hold.—*Examples:* (a) A single point p of a simplex $[\sigma]$ represents a deformation retract of $[\sigma]$.—(b) A single point $p \in S^1$ of a 1-sphere represents a retract of S^1, but not a deformation retract of S^1.

D. Homotopy types

2.9 Definition: Two spaces R and S are said to be of the *same homotopy type* if there exist mappings $f: R \to S$ and $g: S \to R$ such that $g \circ f \simeq i_R$ and $f \circ g \simeq i_S$.

Homeomorphic spaces surely belong to the same homotopy type. On the other hand, a simplex $[\sigma^n]$ and a space consisting of a single point represent two nonhomeomorphic spaces of the same homotopy type.— If R and S are of the same homotopy type, and also S and T, then, as one can easily verify, also R and T are of the same homotopy type. We can therefore speak of classes of spaces of the same homotopy type. Each such class decomposes into well-defined space types; the partitioning of topological spaces into homotopy types is a coarsening of the partitioning into space types.

The problem of homotopy types demands a decision concerning whether two prescribed spaces do or do not belong to the same homotopy type, and moreover it demands the exhibition of all homotopy types. There are available at this time several methods of solving this problem, which to be sure are algebraically very complicated and lie outside the framework of this little volume. In any case, the problem of homotopy types appears to be essentially simpler than the homeomorphism problem. The majority of all the presently known invariants of topological spaces, in particular all the properties of spaces treated in this volume, are at the same time invariants of the homotopy type. There are only a few methods which can serve to resolve a homotopy type into the various space types contained in it.—We now turn to a detailed theory of polyhedra. In §23, we shall come back to the problems and examples introduced up to this point from the viewpoint then reached.

PART I
SIMPLICIAL COMPLEXES

Chapter 1

GEOMETRY OF SIMPLICIAL COMPLEXES

§ 3. Hulls and Stars

We first give several additions to the theory of simplicial complexes discussed in Volume I, §§30, 31.

3.1 Definition: Let N be an arbitrary subset of a simplicial complex K, i.e., an arbitrary set of simplexes of K. The set of all (proper and improper) faces of simplexes of N is called the *hull* of N; it is denoted by Hu N. The set of all simplexes of K which have a simplex of N as a (proper or improper) face is called the *star of* N, more specifically, the *star of* N *in* K; it is denoted by St_K N.

The reference to K in the concept St N can be important if besides K still other complexes containing N play a role; such a reference to K is not necessary for the concept Hu N inasmuch as the faces of the simplexes of N are uniquely determined independently of K by N.—We have $N \subset$ Hu N, $N \subset$ St N. Hu N is a subcomplex of K, St N is in general not.—In particular, this explains the concepts Hu σ and St σ for a single simplex σ of K. (As always in the sequel, we identify σ with the set N if N contains only σ.) For a single vertex $\mathfrak{p}$ we have Hu $\mathfrak{p} = \mathfrak{p}$. The vertex stars St $\mathfrak{p}$ play a special role in the sequel; if we denote as in Volume I, Definition 30.3, by $|N|$ the body of N, that is, the point set union of all the open simplexes of N, then $|St\ \mathfrak{p}| =$ st $\mathfrak{p}$ in the sense of our earlier Definition 30.5 from Volume I.

3.2 Definition: A set N of simplexes of a complex is said to be *closed* if N = Hu N; it is said to be *open*, more specifically, *open in* K, if St N = St_KN = N.

These concepts belonging to the purely combinatorial theory of simplicial complexes are not to be confused with the concepts "open" and "closed" for subsets of a topological space $\mathscr{H}$. The following relationship holds between the two: If N is closed, then | N | considered as a subset of a topological space $\mathscr{H}$ = | K | is closed; if N is open in K, then | N | is open in $\mathscr{H}$. The first is valid because | N | is built up of a finite number of closed simplexes, the second follows by a transition to the closed set K − N.—If N is closed or open, then K − N is open or closed respectively.—The open subsets of K are a special case of the so-called subcomplexes of K, which are defined as follows:

3.3 Definition: A subset N of simplexes of a complex K is called a *subcomplex* of K if N is open in Hu N.* Hu N − N = Bd N is called the *exterior boundary*, or briefly the *boundary*, of the subcomplex N.

A subcomplex N is open in the subcomplex K_0 = Hu N of K. Conversely, if a subset N of K is open in a subcomplex K_0 of K, then N is *a fortiori* open in Hu N ⊂ K_0; it is therefore a subcomplex of K. A subcomplex N of K can thus also be defined as a subset of K, which is open in a suitable subcomplex K_0 of K.—Bd N is by definition closed, and therefore it is a subcomplex of K.

Examples: (a) Every open and every closed subset of K is a subcomplex of K.—(b) A single simplex σ of K represents a subcomplex which is open in Hu σ but in general not in K.—(c) The star St σ of a simplex σ of K is a subcomplex of K; it is open in K.—Further examples are given in the immediately following sections.

3.4 Theorem: *If* σ_i $(i = 1, \ldots, k)$ *are arbitrary simplexes of a*

* *Translator's note:* This is not the usual definition of a subcomplex. See, for instance, P. S. Aleksandrov, *Combinational Topology*, Graylock, Vol. 1, p. 71. Our definition includes both open and closed subcomplexes.

complex K, *then the intersection* $\cap$ St $\sigma_i \neq \emptyset$ *if and only if all these simplexes are faces of the same simplex.*

Proof: If σ is a simplex of the intersection, then every σ_i is a side of σ. Conversely, if all σ_i are faces of a simplex σ, then σ belongs to all the St σ_i.

For all vertex stars, the following theorem holds.

3.5 Theorem: *The arbitrarily chosen vertex stars* St $\mathfrak{p}_i$ $(i = 1, \ldots, k)$ *of a complex* K *have intersection* $\neq \emptyset$ *if and only if the* $\mathfrak{p}_i$ *are vertices of a simplex; in this case, if* σ *is a simplex spanned by the* $\mathfrak{p}_i$, *then* $\cap$ St $\mathfrak{p}_i$ = St σ.

The first part of this theorem is simply a special case of the preceding theorem. The second part follows immediately from the proof of the latter.—Compare also the earlier Theorem 30.6 in Volume I.

3.6 Definition: A complex K is called *connected* if any two of its vertices can be joined by means of a polygonal arc in K.

In a simplicial complex K, the set of all vertices decomposes uniquely into a finite number of classes having vertices which are joinable with one another by means of finite polygonal arcs. A simplex of K has all its vertices in the same class. If one combines for instance all simplexes of K which have their vertices in one of these classes, then one obviously obtains a connected subcomplex of K, the so-called *components*, or more specifically the *connectivity components*, of K. Therefore the following theorem holds.

3.7 Theorem: *A simplicial complex* K *decomposes uniquely into a finite number of connected subcomplexes, the components of* K.

3.8 Theorem: *A polyhedron* $R = |\mathsf{K}|$ *is connected if and only if* K *is connected.*

Here, the word "connected" in the first occurrence is to be thought of in the sense of set-theoretic topology (see Volume I, §7) and in the second occurrence in the sense of our last definition.

Proof: That for connected K also | K | is connected follows from Volume I, Theorem 7.5, if, starting from a vertex, we adjoin one after the other the remaining closed simplexes. The converse is clear immediately for every realization of K.

§ 4. Barycentric Stars

The content of this section will be first used in Chapter 8 and it can therefore be postponed until the reading of that chapter.

We recall the normal subdivision K′ of a simplicial complex K which we discussed in Volume I, §31. The simplexes of K′ are of the form $\sigma' = \sigma'^q = \hat{\sigma}_0 \ldots \hat{\sigma}_q$, where $\hat{\sigma}_i$ is the center of gravity of the simplex σ_i of K ($i = 0, \ldots, q$) and $\sigma_0 \prec \ldots \prec \sigma_q$ is an incidence sequence. We call $\hat{\sigma}_0$ the lowest and $\hat{\sigma}_q$ the highest vertex of σ'.— Each open simplex (σ) of K is subdivided, in the transition to K′, into those open simplexes of K′ which have $\hat{\sigma}$ as the highest vertex. Analogous to this remark we now make the following definition.

4.1 Definition: Let σ be a simplex of a simplicial complex K. The *barycentric star*, or briefly the *barystar* of σ, $\tau = \tau(\sigma)$, consists of all those simplexes σ' of K′ which have $\hat{\sigma}$ as the lowest vertex. The center of gravity of σ, $\hat{\sigma}$, is also called the center of gravity $\hat{\sigma} = \hat{\tau}$ of τ; if, in particular, $\sigma = \mathfrak{p}$ a vertex of K, then $\tau(\mathfrak{p})$ is called the *barycentric vertex star* of K. Hu τ is called a *closed* barystar.

A q-dimensional simplex from $\tau = \tau(\sigma)$ is a simplex from K′ of the form $\sigma' = \hat{\sigma}\hat{\sigma}_1 \ldots \hat{\sigma}_q$ with $\sigma \prec \sigma_1 \prec \ldots \prec \sigma_q$. Every simplex of K′ occurs in a barystar, namely in the barystar of its lowest vertex; but it also occurs only in this barystar and in no other. The totality of barystars of K therefore form a partition (see the Index of Volume I) of K′. Every simplex of K′ and every point of | K | has a well-defined barystar as carrier.—A face simplex of $\tau = \tau(\sigma)$ is either of the form (a): $\hat{\sigma}\hat{\sigma}_{i1} \ldots \hat{\sigma}_{ip}$ or of the form (b): $\hat{\sigma}_{i1} \ldots \hat{\sigma}_{ip}$, where in both cases $\sigma \prec \sigma_{i1} \prec \ldots \prec \sigma_{ip}$ is a subsequence of $\sigma \prec \sigma_1 \prec \ldots \prec \sigma_q$. The faces of the form (a) themselves belong to τ, those of the form (b) must be adjoined to τ if one wishes to obtain Hu τ. Since no face of the form (b) has a simplex of τ as a face, τ is open in Hu τ. A barystar is therefore a subcomplex of K′. — We also write the boundary (defined by Definition 3.3) briefly as Bd $\tau = \dot{\tau}$. Then $\dot{\tau}$ consists precisely of the simplexes of K′ of the form (b). These simplexes belong to the barystars $\tau_{i1} = \tau(\sigma_{i1})$, and, conversely, all simplexes of τ_{i1} are of the form (b). Therefore $\dot{\tau}$ is the union of all $\tau_i = \tau(\sigma_i)$

with $\sigma \prec \sigma_i$. We write the fact that τ_i belongs to $\dot{\tau}$ in the form $\tau_i \prec \tau$. We can then say: If $\sigma_1 \prec \sigma_2$ and if $\tau_1 = \tau(\sigma_1)$ and $\tau_2 = \tau(\sigma_2)$ are the corresponding barystars, then $\tau_2 \prec \tau_1$; conversely, it also follows from $\tau_2 \prec \tau_1$ that $\sigma_1 \prec \sigma_2$. We summarize all this in the following theorem.

4.2 Theorem: *The barystars of a simplicial complex* K *form a partition of the normal subdivision* K′ *of* K *into subcomplexes. The simplexes* σ *of* K *and the barystars* τ *correspond in one-to-one fashion, if* $\sigma_1 \prec \sigma_2$, *then* $\tau_2 \prec \tau_1$, *and conversely.*

Since every simplex σ of K is either a vertex or it possesses a vertex as a proper face, we have, conversely, that every barystar is either a barycentric vertex star or it lies on the boundary of a barycentric vertex star. The closed barycentric vertex stars Hu $\tau(\mathfrak{p})$ therefore comprise all barystars and thus all simplexes of K′. The corresponding point sets $|$ Hu $\tau(\mathfrak{p})$ $|$ which are closed in $|$ K $|$ accordingly cover $|$ K $|$. Several closed barycentric vertex stars have non-empty intersection if and only if the corresponding vertices span a simplex. This fact can be expressed as follows:

4.3 Theorem: *The closed barycentric vertex stars* $|$ Hu $\tau(\mathfrak{p})$ $|$ *of* K *form a closed covering of* $|$ K $|$ *whose nerve is represented by* K.

If we think of a simplex $\sigma \in$ K, as described in the introductory section, as the union of K′-simplexes, then the intersection of σ_1 and $\tau(\sigma_2)$ obviously consists of those K′-simplexes which have $\hat{\sigma}_2$ as the lowest vertex and $\hat{\sigma}_1$ as the highest vertex. The simplex σ and $\tau(\sigma)$ have as intersection only the common center of gravity $\hat{\sigma} = \hat{\tau}$. In general, the following theorem holds.

4.4 Theorem: *A simplex* σ_1 *from* K, *considered as the union of* K′-*simplexes, and the barystar* $\tau(\sigma_2)$ *of a simplex* σ_2 *of* K *have an intersection* $\neq \emptyset$ *if and only if* $\sigma_2 = \sigma_1$ *or* $\sigma_2 \prec \sigma_1$.

Without any greater difficulty, one also recognizes the validity of the following theorem.

4.5 Theorem: *A closed simplex* $[\sigma_1]$, *considered as a union of* K′-*simplexes, and the closed barystar* $[\tau(\sigma_2)]$ *of a simplex* σ_2 *of* K *have an intersection* $\neq \emptyset$ *if and only if* $\sigma_2 = \sigma_1$ *or* $\sigma_2 \prec \sigma_1$.

§ 5. Simplicial Mappings

Let $\sigma^r = \mathfrak{p}_0 \ldots \mathfrak{p}_r$ and $\tau^s = \mathfrak{q}_0 \ldots \mathfrak{q}_s$ be an r-dimensional simplex and an s-dimensional simplex respectively. Let there be assigned to each vertex $\mathfrak{p}_i$ ($i = 0, \ldots, r$) of σ^r some vertex $\varphi(\mathfrak{p}_i) = \mathfrak{q}_j$ (j from the sequence $0, \ldots, s$) of τ^s, and to each point $\mathfrak{x}$ of the closed simplex $[\sigma^r]$ having the representation $\mathfrak{x} = \sum \lambda_i \mathfrak{p}_i$ with $\sum \lambda_i = 1$, $\lambda_i \geqq 0$, let there be assigned the point $\varphi(\mathfrak{x}) = \mathfrak{y} \in [\tau^s]$ having the representation $\mathfrak{y} = \sum \lambda_i \varphi(\mathfrak{p}_i)$. One recognizes that $\mathfrak{y}$ actually lies in $[\tau^s]$ if one combines the summands with distinct i but equal $\varphi(\mathfrak{p}_i)$ so that the form $\mathfrak{y} = \sum \mu_j \mathfrak{q}_j$ results, where then $\mu_j \geqq 0$, $\sum \mu_j = 1$. The mapping φ: $[\sigma^r] \to [\tau^s]$ so defined is evidently continuous; it is called an *affine mapping of* σ^r *into* τ^s. It is already completely determined by the values $\varphi(\mathfrak{p}_i)$.—If the $\varphi(\mathfrak{p}_i)$ are not different from one another, as is the case for example when $s < r$, then φ is called a *degenerate mapping*.

5.1 Theorem: *Under an affine mapping of the simplex* σ^r *into the simplex* τ^s *the interior points of* σ^r *map into interior points of image simplexes (which properly or improperly bound* τ^s*) of* σ^r.

The interior points of σ^r are namely defined by $\lambda_i > 0$ for all i; as one sees, they go over into points $\mathfrak{y}$ with $\mu_j > 0$ for all j.—Moreover, in no way do the boundary points of σ^r always go over into boundary points of τ^s.

5.2 Definition: A mapping φ: $|\mathsf{K}| \to |\Lambda|$ is called a *simplicial* mapping if for every simplex σ of K it is true that: (1) The vertices of σ are mapped into the vertices of a simplex of Λ; (2) the interior points of σ are mapped according to the affine mapping determined by (1).

A simplicial mapping φ of $|\mathsf{K}|$ into $|\Lambda|$ is a continuous mapping. It suffices to see this for simplicial complexes K and Λ in the unit realization (cf. Volume I, §30) inasmuch as the transition from the unit realization of a polyhedron to an arbitrary representation is one to one and continuous. In the unit realization the barycentric and cartesian coordinates coincide and φ is expressed, as one deduces from the introductory remarks of this section, linearly in these coordinates which guarantees the continuity of φ.—The definition of a simplicial mapping φ of a polyhedron R into a polyhedron S

implies reference to a simplicial decomposition K of R and a simplicial decomposition Λ of S; in general, φ is not simplicial with respect to other simplicial decompositions of R and S. In order to emphasize this in the simplest possible way, we write simplicial mappings in the sequel in the form $\varphi\colon \mathsf{K} \to \Lambda$ and take into consideration the innate inconsequence of the relation (φ maps $|\mathsf{K}|$ into $|\Lambda|$ and not K into Λ).—The following two theorems which will subsequently be used a great deal follow immediately so that a particular proof is superfluous.

5.3 Theorem: *If one assigns to each vertex* $\mathfrak{p}_i$ *of a simplicial complex* K *a vertex* $\mathfrak{q}_j = \varphi(\mathfrak{p}_i)$ *of a simplicial complex* Λ *in such a way that to vertices of a simplex of* K *there always correspond vertices of a simplex of* Λ, *then this determines uniquely a simplicial mapping.*

5.4 Theorem: *If* $\varphi\colon \mathsf{K} \to \Lambda$ *and* $\psi\colon \Lambda \to \mathsf{M}$ *are simplicial mappings, then* $\psi \circ \varphi\colon \mathsf{K} \to \mathsf{M}$ *is also a simplicial mapping.*

We give a simple example of a simplicial mapping which will at the same time be especially important for our later development, namely the *displacement* $\eta\colon \mathsf{K}' \to \mathsf{K}$ *of* K' *toward* K, where K is a given simplicial complex and K' is an arbitrary subdivision of K. The simplicial mapping η is defined as follows: To each vertex $\mathfrak{p}'$ of K' one assigns an arbitrarily chosen vertex $\mathfrak{p} = \eta(\mathfrak{p}')$ of the carrier simplex of $\mathfrak{p}'$ in K. We must show that the vertices $\mathfrak{p}'_i$ of an arbitrary q-dimensional simplex σ' of K' ($i = 0, \ldots, q$) go over under η into vertices of a simplex. Now, by the definition of a partition (cf. Volume I, Definition 31.2), the open simplex (σ') is contained entirely in an open simplex (σ) of K. It follows from $(\sigma') \subset (\sigma)$ that for the closed simplexes we have $[\sigma'] \subset [\sigma]$. Accordingly, we have that $\mathfrak{p}'_i \in [\sigma'] \subset [\sigma]$, the carrier simplexes of the $\mathfrak{p}'_i$ are face simplexes of σ, the $\eta(\mathfrak{p}'_i)$ are therefore vertices of σ, which was to be shown.—Let $\mathfrak{x}$ be an arbitrary point of $|\mathsf{K}'|$, say with the simplex σ' just studied as carrier in K'. Under η, σ' goes over into a face simplex of σ, on which therefore $\eta(\mathfrak{x})$ also lies. The points $\mathfrak{x}$ and $\eta(\mathfrak{x})$ can be joined rectilinearly in the interior of $[\sigma]$. If we let $\mathfrak{x}$ range proportionally with a parameter t, $0 \leqq t \leqq 1$, to $\eta(\mathfrak{x})$, then we obtain a homotopy which joins the identity mapping of $|\mathsf{K}|$ with η. The displacement η is therefore a deformation of the identity.—We should make clear the concept of the displacement η: $\mathsf{K}' \to \mathsf{K}$ especially in the case when K' is a normal subdivision of K. We shall generalize this concept in an essential way at the end of this section.

5.5 Definition: Let f: $|\mathsf{K}| \to |\Lambda|$ be a continuous mapping. The simplicial mapping φ: $\mathsf{K} \to \Lambda$ is called a *simplicial approximation* of f if for every point $\mathfrak{x} \in |\mathsf{K}|$ either one of the following two equivalent conditions is fulfilled:

(a) $\varphi(\mathfrak{x})$ lies on the closed carrier simplex of $f(\mathfrak{x})$;

(b) $f(\mathfrak{x})$ lies on the star of the carrier simplex of $\varphi(\mathfrak{x})$.

The simplicial mappings form in our development the combinatorial equivalent of continuous mappings.—The equivalence of the conditions (a) and (b) is self-evident.—(a) asserts that $\varphi(\mathfrak{x})$ lies at not too great a distance from $f(\mathfrak{x})$, the distance is at most equal to the largest diameter of the simplexes of Λ.—The following theorem also shows the close relation between f and φ.

5.6 Theorem: *If φ: $\mathsf{K} \to \Lambda$ is a simplicial approximation of the continuous mapping f: $|\mathsf{K}| \to |\Lambda|$, then $\varphi \simeq f$; two simplicial approximations of f are mutually homotopic.*

Proof: We join $f(\mathfrak{x})$ in the interior of its closed carrier simplex rectilinearly with $\varphi(\mathfrak{x})$. On this joining segment, we let $f(\mathfrak{x})$ go over with "constant velocity", proportional to the parameter t, $0 \leqq t \leqq 1$, to $\varphi(\mathfrak{x})$. We thus obtain a joining homotopy between f and φ. The displacement η introduced above is a simplicial approximation of the identity mapping i: $|\mathsf{K}'| \to |\mathsf{K}|$. The second assertion follows directly from the first.

5.7 Theorem: *A simplicial approximation φ of a simplicial mapping f is identical with f.*

Proof: The vertices $\mathfrak{p}$ of K are mapped by f into the vertices $f(\mathfrak{p}) = \mathfrak{q}$ of Λ. The criterion of Definition 5.5 asserts that also $\varphi(\mathfrak{p}) = \mathfrak{q}$; but since φ is uniquely determined by the values $\varphi(\mathfrak{p})$, it follows that $f = \varphi$.

5.8 Theorem: *If $\varphi\colon \mathsf{K} \to \Lambda$ is a simplicial approximation of the continuous mapping $f\colon |\mathsf{K}| \to |\Lambda|$, then the relation*

$$f(\mathrm{st}\,\mathfrak{p}) \subset \mathrm{st}\,\varphi(\mathfrak{p})$$

is valid for every vertex $\mathfrak{p}$ of K.

Naturally, on the left st_{K} is meant whereas on the right st_{Λ} is meant.

Proof: Let $\mathfrak{x}$ be an arbitrary point of st $\mathfrak{p}$. Then $\mathfrak{x}$ is an interior point of a simplex σ with the vertex $\mathfrak{p}$. Thus, $\varphi(\mathfrak{x})$ is an interior point of a simplex τ with the vertex $\varphi(\mathfrak{p})$ (cf. Theorem 5.1). The carrier of $\varphi(\mathfrak{x})$ is τ. By condition (b) of Definition 5.5, $f(\mathfrak{x})$ lies in $|\,\mathrm{St}\,\tau\,|$. Since obviously $|\,\mathrm{St}\,\tau\,| \subset \mathrm{st}\,\varphi(\mathfrak{p})$, it follows that $f(\mathfrak{x}) \in \mathrm{st}\,\varphi(\mathfrak{p})$, which completes the proof of the theorem.—With regard to this theorem, we make the following definition.

5.9 Definition: Let f be a continuous mapping of the polyhedron R into the polyhderon S. We say that the simplical subdivisions K of $R = |\mathsf{K}|$ and Λ of $S = |\Lambda|$ fulfill the *star condition with respect to f* if for each vertex $\mathfrak{p}$ of K there exists a vertex $\mathfrak{q}$ of Λ such that

$$f(\mathrm{st}\,\mathfrak{p}) \subset \mathrm{st}\,\mathfrak{q} \text{ (star condition).}$$

The star condition means roughly speaking that for given f the simplexes of K are small in comparison with the simplexes of Λ.

5.10 Theorem: *If f is a continuous mapping of the polyhedron $R = |\mathsf{K}|$ into the polyhedron $S = |\Lambda|$, then there exists a natural number ν such that the ν-th normal subdivision $\mathsf{K}^{(\nu)}$ of K together with Λ satisfy the star condition with respect to f.*

Proof: The totality of vertex stars st $\mathfrak{q}_j$ of Λ $(j = 1, \ldots, \beta)$ form an open covering of $|\Lambda|$; the totality of $f^{-1}(\mathrm{st}\ q_i)$ therefore form an open covering of $|K|$. Let λ be a Lebesgue number for this covering and let ν be so chosen that the largest simplex diameter of $K^{(\nu)}$ is smaller than $\lambda/2$, which is possible by Theorem 31.6, Volume I. Then for all vertices p_i of K we have $d(\mathrm{st}\ p_i) < \lambda$ $(i = 1, \ldots, \alpha)$. Every star st p_i is therefore contained in a suitable set $f^{-1}(\mathrm{st}\ q_i)$. From this it follows that $f(\mathrm{st}\ p_i) \subset \mathrm{st}\ q_i$, which was to be proved.

5.11 Theorem: *Let* $f\colon |K| \to |\Lambda|$ *be a continuous mapping. If* K *and* Λ *fulfill the star condition with respect to* f *and if one assigns to each vertex* p *of* K *the vertex* $q = \varphi(p)$ *of* Λ *specified in the star condition, then* φ *defines a simplicial approximation* $\varphi\colon K \to \Lambda$ *of* f.

Proof: Let $\sigma = p_0 \ldots p_q$ be an arbitrary simplex of K and let $x \in (\sigma)$ be an arbitrary point in this simplex. Then $x \in \mathrm{st}\ p_i$ for $i = 0, \ldots, q$. According to the star condition, it follows that $f(x) \in \mathrm{st}\ \varphi(p_i)$. The stars st $\varphi(p_i)$, therefore, have a non-empty intersection, from which it follows by virtue of Theorem 3.4 that the $\varphi(p_i)$ are vertices of a simplex of Λ. Thus φ defines in every case a simplicial mapping; we denote it also by φ.—Now, more specifically, let $\tau = \varphi(\sigma)$ be the image simplex of σ generated by the $\varphi(p_i)$. The carrier of $\varphi(x)$ is τ. The image $f(x)$ belongs to the intersection $\bigcap \mathrm{st}\ \varphi(p_i)$, which by virture of Theorem 3.5 is equal to st τ. Therefore $f(x)$ belongs to the star of the carrier of $\varphi(x)$. This is exactly the condition (b) of Theorem 3.4. This completes the proof of the theorem.

A combination of the last two theorems now yields immediately the following approximation theorem which is the principal goal of this section.

5.12 Theorem: *For each continuous mapping* $f\colon |K| \to |\Lambda|$ *there exists a simplicial approximation* φ *of* f *which is a simplicial mapping of a suitable* ν*-th normal subdivision* $K^{(\nu)}$ *of* K *into* Λ.

The preceding investigations of simplicial approximations of continuous mappings f are also of interest in the special case when f

is the identity mapping of the space R. Accordingly, if—with a slight modification of notation—$R = |K_1| = |K_2|$ and $\eta: K_1 \to K_2$ is a simplicial approximation of the identity $i: R \to R$, then η is called a *displacement of* K_1 *to* K_2. This definition evidently comprises the current explanation of this concept stated immediately after Theorem 5.4. According to Theorem 5.8, K_1 and K_2 satisfy the star condition with respect to i, which in this case asserts: *For every vertex* $p_1 \in K_1$ *there exists a vertex* $p_2 \in K_2$ *with* st $p_1 \subset$ st p_2. We say briefly without reference to i that K_1 and K_2 satisfy the star condition or still more briefly that K_1 is small in comparison with K_2, in symbols $K_1 < K_2$. Then the following holds: $K_1 < K_2$ and $K_2 < K_3$ imply that $K_1 < K_3$. Theorem 5.10 asserts in this case that if K and Λ are two simplical decompositions of $R = |K| = |\Lambda|$ then there always exists a natural number ν with $K^{(\nu)} < \Lambda$.

§ 6. Neighboring Mappings

6.1 Theorem: *For every polyhedron S there exists a number $\vartheta > 0$—a so-called homotopy bound for S—such that every two mappings f, g: $R \to S$ of an arbitrary polyhedron R into S with $d(f, g) < \vartheta$ are homotopic and that f and g possess a common simplicial approximation.*

In this connection, the distance is understood to be $d(f, g) = \text{Max}\ d(f(x), g(x))$ taken over all $x \in R$ as in §25, Volume I.—The first assertion of the theorem is of special geometric interest, the second assertion will be important for our combinatorial investigation of the homotopy concept.—If ϑ is a homotopy bound for S then so is every ϑ' with $0 < \vartheta' < \vartheta$.—The question of the upper limit of all homotopy bounds of a polyhedron is difficult to answer.

Proof: Let $R = |K|$, $S = |\Lambda|$. Let λ be a Lebesgue number of the covering of $|\Lambda|$ by the vertex stars st q_i ($i = 1, \ldots, \beta_0$). Let ϑ equal $\lambda/3$. We form the sets

$$U_i = \{\mathfrak{x} \mid \mathfrak{x} \in |\Lambda|; d(\mathfrak{x}, |\Lambda| - \text{st}\ q_i) > \vartheta\}.$$

These U_i are open and they are contained in st q_i. No $q_j \neq q_i$ is contained in U_i.—We first show that the U_i also form a covering for $|\Lambda|$. Namely, let $\mathfrak{y}$ be an arbitrary point of $|\Lambda|$. Let

$$A = \{\mathfrak{x} \mid \mathfrak{x} \in |\Lambda|;\, d(\mathfrak{x}, \mathfrak{y}) \leqq \vartheta\}.$$

A has a diameter $d(A) \leqq 2\vartheta < 3\vartheta = \lambda$; therefore, as a consequence of the properties of λ, A is contained in one of the stars st $\mathfrak{q}_i$. Furthermore,

$$d(\mathfrak{y}, |\Lambda| - \text{st } \mathfrak{q}_i) > \vartheta,$$

as follows first for each point of $|\Lambda|$ − st $\mathfrak{q}_i$ instead of this set and then also for this compact set itself (according to Volume I, 24.1). Therefore, $\mathfrak{y} \in U_i$. Thus the U_i cover $|\Lambda|$. Moreover, it now follows that $\mathfrak{q}_i \in U_i$ which was evident directly: U_i is a neighborhood of $\mathfrak{q}_i$.—The sets $f^{-1}(U_i)$ form a covering of $|\mathsf{K}|$. Let λ' be a corresponding Lebesgue number and let K' be chosen as an iterated normal subdivision of K so that the diameter of every simplex of K' is smaller than $\lambda'/2$. Then $d(\text{st } \mathfrak{p}_j') < \lambda'$ for every vertex $\mathfrak{p}_j'$ of K'. According to the properties of λ', there exists for j an index i with st $\mathfrak{p}_j' \subset f^{-1}(U_i)$, and hence, $f(\text{st } \mathfrak{p}_j') \subset U_i \subset$ st $\mathfrak{q}_i$. We assert that also $g(\text{st } \mathfrak{p}_j') \subset$ st $\mathfrak{q}_i$. When this is proved, the ordering of $\mathfrak{q}_i$ to $\mathfrak{p}_j'$ yields, according to Theorem 5.11, a common simplicial approximation φ of f and g, and it follows that $f \simeq \varphi$, $g \simeq \varphi$, and therefore $f \simeq g$, which completes the proof of the theorem.—Therefore, let $\mathfrak{x}$ be an arbitrary fixed point from st $\mathfrak{p}_j'$. By virtue of the second triangle inequality (cf. § 1, Volume I) for an arbitrary point $\mathfrak{x} \in |\Lambda|$ − st $\mathfrak{q}_i$ we have

$$d(\mathfrak{x}, g(\mathfrak{x})) \geqq |\, d(\mathfrak{x}, f(\mathfrak{x})) - d(f(\mathfrak{x}), g(\mathfrak{x}))\, |.$$

Since $f(\mathfrak{x}) \in U_i$, we have, by the definition of U_i, that the first summand on the right is $>\vartheta$ and the second summand is in absolute value $<\vartheta$ so that $d(\mathfrak{x}, g(\mathfrak{x})) > 0$. Since the set $|\Lambda|$ − st $\mathfrak{q}_i$ is compact, it follows that $d(g(\mathfrak{x}), |\Lambda| - \text{st } \mathfrak{q}_i) > 0$; hence, $g(\mathfrak{x}) \in$ st $\mathfrak{q}_i$ and $g(\text{st } \mathfrak{p}_j') \subset$ st $\mathfrak{q}_i$, which was still to be shown.—It follows, in particular, from the theorem just proved that if $f\colon R \to R$ is a mapping of R into itself which is sufficiently close to the identity, for which, therefore, $d(\mathfrak{x}, f(\mathfrak{x})) < \epsilon$ holds for suitably small ϵ for all $\mathfrak{x} \in R$, then f is homotopic to the identity mapping.—Moreover, the preceding Theorem 5.12 is a special case of the theorem just proved, which results when one sets $f = g$.

We come now to the concept which will yield a combinatorial equivalent of the homotopy concept.

6.2 Definition: Two simplicial mappings φ, ψ: $\mathsf{K} \to \Lambda$ are said to be *neighboring* if for every simplex $\sigma \in \mathsf{K}$ the two simplexes $\varphi(\sigma)$ and $\psi(\sigma)$ are faces of the same simplex $\tau \in \Lambda$.

6.3 Theorem: *Two simplicial approximations* φ, ψ: $\mathsf{K} \to \Lambda$ *of a continuous mapping* f: $|\mathsf{K}| \to |\Lambda|$ *are neighboring.*

Proof: Let $\sigma \in \mathsf{K}$, $\mathfrak{x}$ an interior point of σ and $[\tau]$ the closed carrier simplex of $f(\mathfrak{x})$. We shall show that $\varphi(\sigma)$ and $\psi(\sigma)$ are faces of τ. Namely, $\varphi(\mathfrak{x})$ lies (according to Definition 5.5(a)) on $[\tau]$, $\varphi(\mathfrak{x})$ is an interior point of $\varphi(\sigma)$ (Theorem 5.1), $\varphi(\sigma)$ therefore lies on $[\tau]$. The same holds for $\psi(\sigma)$, which completes the proof of the theorem.

The relation of being neighboring between two simplicial mappings is obviously symmetric but it is not transitive (draw a two-dimensional counterexample!). It gives rise nonetheless to the following transitive relation.

6.4 Definition: Two simplicial mappings φ, ψ: $\mathsf{K} \to \Lambda$ are said to *belong to the same neighborhood class* if there exists a chain of simplicial mappings φ_i: $\mathsf{K} \to \Lambda$ $(i = 0, 1, \ldots, k)$ such that $\varphi = \varphi_0$, $\psi = \varphi_k$ and φ_i is neighboring to φ_{i-1}.

One sees immediately that if φ: $\mathsf{K} \to \Lambda$ as well as ψ: $\mathsf{K} \to \Lambda$ are in the same neighborhood class as χ: $\mathsf{K} \to \Lambda$, then φ and ψ are also in the same neighborhood class.

6.5 Theorem: *If the two continuous mappings* f, g: $|\mathsf{K}| \to |\Lambda|$ *are homotopic, then there exist simplicial approximations* φ *of* f *and* ψ *of* g *which are simplicial mappings of a suitable* ν*-fold normal subdivision* $\mathsf{K}^{(\nu)}$ *of* K *into* Λ *and lie in the same neighborhood class.*

Proof: Let a homotopy joining f and g be given in the form F: $|\mathsf{K}| \times I \to |\Lambda|$, where I denotes the unit interval $0 \leqq t \leqq 1$ and $F(\mathfrak{x}, t) = \mathfrak{y}$ with $\mathfrak{x} \in |\mathsf{K}|$, $\mathfrak{y} \in |\Lambda|$ and $F(\mathfrak{x}, 0) = f(\mathfrak{x})$, $F(\mathfrak{x}, 1) = g(\mathfrak{x})$, $F(\mathfrak{x}, t) = f_t(\mathfrak{x}) = f_t$. But $F(\mathfrak{x}, t)$ is uniformly continuous (Volume I, Theorem 25.1) from which one deduces the following: For every $\epsilon > 0$ there exists a $\delta > 0$ such that

$$d(F(\mathfrak{x}, t), F(\mathfrak{x}, t')) < \epsilon$$

provided $|t - t'| < \delta$ for *all* $\mathfrak{x}$. This fact can also be written in the form $d(f_t, f_{t'}) < \epsilon$ provided $|t - t'| < \delta$.—Now let $\epsilon = \vartheta > 0$ be a homotopy bound of $|\Lambda|$. We determine numbers $0 = t_0 < t_1 < \ldots < t_k = 1$ such that $|t_i - t_{i-1}| < \delta$ $(i = 1, \ldots, k)$. Then $d(f_{t_i}, f_{t_{i-1}}) < \vartheta$, and f_{t_i} and $f_{t_{i-1}}$ possess, by virtue of Theorem 6.1, a common simplicial approximation φ_i: $\mathsf{K}^{(\nu_i)} \to \Lambda$, where $\mathsf{K}^{(\nu_i)}$ is a suitable iterated normal subdivision of K Let ν be the largest of the numbers ν_i. Then, by virtue of Theorem 6.1, f_{t_i} and $f_{t_{i-1}}$ also possess a common simplicial approximation $\bar{\varphi}_i$: $\mathsf{K}^{(\nu_i)} \to \Lambda$, where $\mathsf{K}^{(\nu_i)}$ is a suitable iterated normal subdivision of K. Then f_{t_i} and $f_{t_{i-1}}$ also possess by virtue of Theorem 6.1 a common simplicial approximation φ_i: $\mathsf{K}^{(\nu)} \to \Lambda$. Now φ_1 is a simplicial approximation of $f_{t_0} = f$, φ_k is a simplicial approximation of $f_{t_k} = g$, two successive simplicial approximations φ_1 and φ_{i+1} are, as simplicial approximations of the same continuous mapping f_{t_i}, mutually neighboring by virtue of Theorem 6.3. This completes the proof of the theorem.

Chapter 2

HOMOLOGY GROUPS AND COHOMOLOGY GROUPS

§ 7. Orientation. Incidence Numbers

The $(q + 1)!$ arrangements of the $q + 1$ vertices of a q-dimensional simplex σ^q with $q > 0$ decompose into two classes: Arrangements from the same class arise from one another by means of an even permutation, and arrangements from different classes arise from one another by means of an odd permutation. These two classes are called the two *orientations* of the simplex σ^q. A simplex σ^q together with a prescribed orientation is called an *oriented simplex*; if $\mathfrak{p}_0, \ldots, \mathfrak{p}_q$ are the vertices of σ^q in an arrangement of prescribed orientation, then we denote the oriented simplex by $\langle \mathfrak{p}_0 \ldots \mathfrak{p}_q \rangle$. Therefore, for each simplex σ^q with $q > 0$, there exist exactly two corresponding oriented simplexes. In the case $q = 1$, for example, the two oriented simplexes $\langle \mathfrak{p}_0\mathfrak{p}_1 \rangle$ and $\langle \mathfrak{p}_1\mathfrak{p}_0 \rangle$ correspond to the segment $\mathfrak{p}_0\mathfrak{p}_1$. We can picture $\langle \mathfrak{p}_0\mathfrak{p}_1 \rangle$ as a segment which is "directed" from $\mathfrak{p}_0$ to $\mathfrak{p}_1$. For $q = 2$, the two orientations $\langle \mathfrak{p}_0\mathfrak{p}_1\mathfrak{p}_2 \rangle = \langle \mathfrak{p}_1\mathfrak{p}_2\mathfrak{p}_0 \rangle = \langle \mathfrak{p}_2\mathfrak{p}_0\mathfrak{p}_1 \rangle$ and $\langle \mathfrak{p}_0\mathfrak{p}_2\mathfrak{p}_1 \rangle = \langle \mathfrak{p}_2\mathfrak{p}_1\mathfrak{p}_0 \rangle = \langle \mathfrak{p}_1\mathfrak{p}_0\mathfrak{p}_2 \rangle$ correspond to the triangle $\mathfrak{p}_0\mathfrak{p}_1\mathfrak{p}_2$. We can interpret these oriented triangles as triangles having one or the other "sense of traversal." Analogously, we can recognize for higher q a spatial "sense of rotation" in the definition. For $q = 0$, of course, there corresponds to the point $\mathfrak{p}_0$ only one arrangement, namely $\langle \mathfrak{p}_0 \rangle$.

The two orientations of a simplex σ^q with $q > 0$ are entirely equivalent, neither can be distinguished from the other in a natural way. But if we choose arbitrarily one of the two orientations and denote it as the positive orientation and the other as the negative orientation, then we say that the simplex σ^q has been oriented. Of the two oriented simplexes belonging to σ^q, denote the one with the positive orientation by $s_q = +s_q$ and the one with the negative orientation by $-s_q$. In an analogous manner, for the case $q = 0$, assign to a vertex $\mathfrak{p}$ the two oriented 0-simplexes $+\langle \mathfrak{p} \rangle$ and $-\langle \mathfrak{p} \rangle$, the positively oriented point and the negatively oriented point. Whereas for $q > 0$, we must choose which of the two oriented simplexes corresponding to σ^q should be the positive one $+s_q$ and which the negative one $-s_q$, for $q = 0$, we always choose $+\langle \mathfrak{p} \rangle$ as the positive and $-\langle \mathfrak{p} \rangle$ as the negative point.

The $(q - 1)$-dimensional faces of a q-dimensional oriented simplex s_q with $q > 0$ can be oriented in a natural way as follows. If $s_q = \langle \mathfrak{p}_0 \ . \ . \ . \ \mathfrak{p}_q \rangle$, then the so-called *induced orientations* of the $(q - 1)$-dimensional face simplexes are defined by

$$s^i_{q-1} = (-1)^i \langle \mathfrak{p}_0 \ldots \hat{\mathfrak{p}}_i \ldots \mathfrak{p}_q \rangle,$$

where the circumflex means that the corresponding vertex is to be omitted. Of course, this definition is meaningful only if it turns out to be independent of the particular representation of s_q, i.e., if it leads to the same s^i_{q-1} when one performs an even permutation in the representation $s_q = \langle \mathfrak{p}_0 \ . \ . \ . \ \mathfrak{p}_q \rangle$. In fact, if one performs an interchange of neighboring vertices in $\langle \mathfrak{p}_0 \ . \ . \ . \ \mathfrak{p}_q \rangle$, then in the definition of s^i_{q-1} either the sign $(-1)^i$ changes if one of the vertices has the index i or the sign of $\langle \mathfrak{p}_0 \ . \ . \ . \ \hat{\mathfrak{p}}_i \ . \ . \ . \ \mathfrak{p}_q \rangle$ changes if not. In every case, s^i_{q-1} changes its sign. But since every even permutation can be obtained by means of an even number of adjacent interchanges, the correctness of our definition of the induced orientation is shown. We say that the simplexes oriented in this way have the incidence numbers $[s_q \colon s^i_{q-1}] = +1$, whereas $[-s_q \colon s^i_{q-1}] = [s_q \colon -s^i_{q-1}] = -1$ and $[-s_q \colon -s^i_{q-1}] = +1$; for non-incident s_q and s_{q-1}, we set $[s_q \colon s_{q-1}] = 0$.—The basis for defining the induced orientation in precisely the given way will show up later; meanwhile we point out only the adjacent figure for the case $q = 2$. The prescribed sense of traversal of the triangle induces in the intuitive sense the given directions of the sides. In the case $q = 1$, the segment $\langle \mathfrak{p}_0\mathfrak{p}_1 \rangle$ directed from $\mathfrak{p}_0$ to $\mathfrak{p}_1$ induces the orientation $+\langle \mathfrak{p}_1 \rangle$ for the end

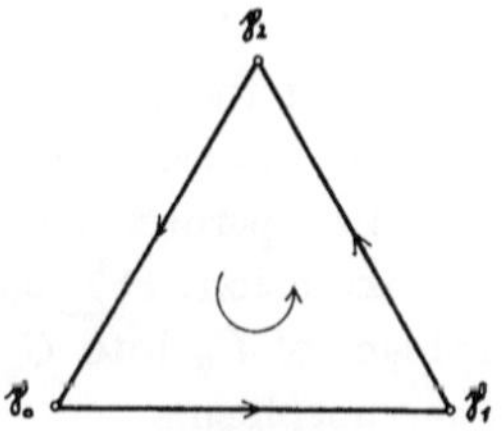

Fig. 2

point and $-\langle \mathfrak{p}_0 \rangle$ for the initial point.—Moreover, it is impossible to induce a natural orientation in the face simplexes of lower than the $(q - 1)$-st dimension of s_q.

§ 8. Homology Groups

Throughout this entire section, let K be an n-dimensional simplicial complex. Let K be oriented, i.e., each simplex of K is oriented in one of the two possible ways, and let the oriented simplexes of dimension q be s_q^i $(i = 1, \ldots, \alpha_q)$. For every fixed q, we form the free module $C_q = C_q(\mathsf{K})$ with the s_q^i as basis elements, i.e., the free abelian group with the s_q^i as generators. Explicitly stated, C_q is the set of all formal sums $c_q = \sum g_i s_q^i$ with g_i an integer, where the sum is taken over the indices $i = 1, \ldots, \alpha_q$, here and in the following. These elements of C_q are called *q-dimensional chains*, or briefly, *q-chains*. They are added and subtracted according to the rule $\sum g_i s_q^i + \sum h_i s_q^i = \sum (g_i + h_i) s_q^i$. For an integer $q < 0$ and for values of q which are larger than the dimension of K, we simply set $C_q = 0$. The free module C_q is called the *q-dimensional chain group*. In particular, the chains s_q^i, the *elementary chains*, and the chains $-s_q^i$, which we have already considered in the preceding section, appear in C_q.

We now introduce the homomorphisms $\partial: C_q \to C_{q-1}$, the *boundary operators*, by first asserting that for oriented simplexes

$$\partial s_q = \partial(\mathfrak{p}_0 \ldots \mathfrak{p}_q) = \sum_{i=0}^{q} (-1)^i \langle \mathfrak{p}_0 \ldots \hat{\mathfrak{p}}_i \ldots \mathfrak{p}_q \rangle. \tag{8.1}$$

Thus, ∂s_q is simply the sum of the $(q-1)$-dimensional boundary simplexes in the orientation induced by s_q. For $q = 0$, we set $\partial\langle\mathfrak{p}\rangle = 0$. We think of this operator as being continued to an entire group C_q by linear extension: $\partial(\sum g_i s_q^i) = \sum g_i(\partial s_q^i)$. Then ∂ is, in fact, a homomorphism of C_q into C_{q-1}. *In toto*, there thus results a sequence of homomorphisms

$$0 \xrightarrow{\partial} C_n \xrightarrow{\partial} \cdots \xrightarrow{\partial} C_q \xrightarrow{\partial} \cdots \xrightarrow{\partial} C_0 \xrightarrow{\partial} 0.$$

We prove now that the application of ∂ twice, which takes a q-dimensional chain into a $(q-2)$-dimensional chain, always yields the zero element. As an operator equation, this is expressed in the form $\partial\partial = \partial^2 = 0$. It suffices to show that $\partial\partial s_q = 0$. In fact, if we apply ∂ once more to (8.1) then only summands of the form $\langle \mathfrak{p}_0 \ldots \hat{\mathfrak{p}}_i \ldots \hat{\mathfrak{p}}_j \ldots \mathfrak{p}_q \rangle$ with certain signs are obtained. Each of these summands arises twice, once $\mathfrak{p}_j$ is omitted first and $\mathfrak{p}_i$ is omitted secondly, and the other time $\mathfrak{p}_i$ is omitted first and $\mathfrak{p}_j$ secondly. One sign is $(-1)^{j+i}$ and the other is $(-1)^{j+i-1}$. The two summands, and hence all summands, therefore, cancel each other out and the result is zero.

The chains $z_q \in C_q$ for which $\partial z_q = 0$ are called *closed chains* or *cycles*. Since $\partial z_q = 0$ and $\partial y_q = 0$ imply that $\partial(z_q - y_q) = 0$, the cycles form a subgroup of C_q, called the *group* $Z_q = Z_q(\mathsf{K})$ *of the q-dimensional cycles of* K. A chain $c_q \in C_q$ that is the boundary of a $(q+1)$-dimensional chain c_{q+1}, i.e., $c_q = \partial c_{q+1}$, is called a *q-dimensional bounding chain* or *boundary chain* of K. Inasmuch as $\partial c_{q+1} - \partial d_{q+1} = \partial(c_{q+1} - d_{q+1})$, these chains form a subgroup $B_q = B_q(\mathsf{K})$ of C_q. If $c_q \in B_q$, i.e., $c_q = \partial c_{q+1}$, then $\partial c_q = \partial^2 c_{q+1} = 0$. Therefore, $c_q \in Z_q$ and B_q is a subgroup of Z_q. The factor group $Z_q/B_q = H_q = H_q(\mathsf{K})$ is called the *q-dimensional homology group* or also the *q-dimensional Betti group* of K. This group will later, in

Chapter 6, be assigned as an invariant to the polyhedron $R = |\mathsf{K}|$ and will, in general, be proved to be one of the most important topological invariants. Meanwhile, we remark only that the elements of $H_q(\mathsf{K})$, the homology classes of K, consist of classes of cycles. Each two cycles of a class are called *homologous*, denoted by $z_q \sim y_q$, and their difference is said to be *homologous to zero*, i.e., $z_q - y_q$ is the boundary of a $(q + 1)$-dimensional chain. In a certain way, H_q measures how many closed but not bounded q-dimensional cycles there are in K.—In the next section, we give examples for the computation of homology groups in simple cases and treat, in Chapter 4, a method for the computation in the general case.—As a factor group of a subgroup of a finitely generated module C_q, H_q itself has a finite generating system. According to the fundamental theorem on finitely generated abelian groups, which we shall prove later (Theorem 16.4), H_q is isomorphic to the direct sum of a number $\wp_q$ of infinite cyclic groups and a number of finite cyclic groups of orders $e_q^1, e_q^2, \ldots, e_q^s$, which form a division chain in this sequence, i.e., $e_q^1 \mid e_q^2, e_q^2 \mid e_q^3, \ldots, e_q^{s-1} \mid e_q^s$. The number $\wp_q$ is called the q-dimensional *Betti number* of K, the natural numbers $e_q^1, \ldots, e_q^s$ are called the q-dimensional *torsion numbers* of K.

Now let $\varphi\colon \mathsf{K} \to \Lambda$ be a given simplicial mapping. We assign to the mapping φ a homomorphism $\varphi.\colon C_q(\mathsf{K}) \to C_q(\Lambda)$. To this end, we set $\varphi.s_q = \varphi.\langle \mathfrak{p}_0 \ldots \mathfrak{p}_q\rangle = \langle \varphi\mathfrak{p}_0 \ldots \varphi\mathfrak{p}_q\rangle$ if the $q + 1$ image points are distinct, i.e., if $\varphi.s_q$ is not degenerate; otherwise, we set $\varphi.\langle \mathfrak{p}_0 \ldots \mathfrak{p}_q\rangle = 0$. By means of a linear extension of $\varphi.$ to $C_q(\mathsf{K})$, in a manner analogous to that above, there results a homomorphism $\varphi.$, as we wished. We assert that $\varphi.$ is boundary-preserving, i.e., in the operator form of writing, $\partial\varphi. = \varphi.\partial$. It suffices to prove this for a simplex s_q, that is, to prove that $\partial\varphi.s_q = \varphi.\partial s_q$. The proof consists of three steps. (1) If $\varphi.s_q$ is not degenerate, then the rule follows simply from

$$\varphi.\partial s_q = \sum (-1)^i \langle \varphi\mathfrak{p}_0 \ldots \widehat{\varphi\mathfrak{p}_i} \ldots \varphi\mathfrak{p}_q\rangle = \partial\varphi.s_q.$$

(2) If precisely one pair of equal vertices occurs in $\varphi.s_q$, then we can assume that they are the first two vertices, that therefore

$\varphi p_0 = \varphi p_1$. Then $\varphi.s_q = 0$; therefore, $\partial\varphi.s_q = 0$, and it remains to prove that $\varphi.\partial s_q = 0$. In fact,

$$\varphi.\partial s_q = \langle \varphi p_1 \varphi p_2 \ldots \varphi p_q \rangle - \langle \varphi p_0 \varphi p_2 \ldots \varphi p_q \rangle = 0$$

because all the other summands vanish. (3) If more than one pair of vertices of $\varphi.s_q$ are equal, then obviously $\partial\varphi.s_q = 0$ and $\varphi.\partial s_q = 0$. That $\varphi.$ preserves boundaries is thus proved. It can also be expressed by means of the following diagram:

$$\begin{array}{ccccccccccc} 0 & \xrightarrow{\partial} & C_n(\mathsf{K}) & \xrightarrow{\partial} & C_{n-1}(\mathsf{K}) & \xrightarrow{\partial} & \ldots & \xrightarrow{\partial} & C_0(\mathsf{K}) & \xrightarrow{\partial} & 0 \\ \downarrow\varphi. & & \downarrow\varphi. & & \downarrow\varphi. & & & & \downarrow\varphi. & & \downarrow\varphi. \\ 0 & \xrightarrow{\partial} & C_n(\Lambda) & \xrightarrow{\partial} & C_{n-1}(\Lambda) & \xrightarrow{\partial} & \ldots & \xrightarrow{\partial} & C_0(\Lambda) & \xrightarrow{\partial} & 0. \end{array}$$

We say that this diagram is commutative on account of the equation $\partial\varphi. = \varphi.\partial$. We can choose two paths from $C_n(\mathsf{K})$ to $C_{n-1}(\Lambda)$, one through $C_{n-1}(\mathsf{K})$ and one through $C_n(\Lambda)$; both yield the same homomorphism $\partial\varphi. = \varphi.\partial$. An important corollary results from the property of boundary preservation. If z_q is a cycle, i.e., $\partial z_q = 0$, then it follows that $\partial\varphi.z_q = \varphi.\partial z_q = 0$, i.e., $\varphi.z_q$ is also a cycle. If c_q is a boundary, $c_q = \partial c_{q+1}$, then $\varphi.c_q = \varphi.\partial c_{q+1} = \partial\varphi.c_{q+1}$ is also a boundary. This means that $\varphi.(Z_q(\mathsf{K})) \subset Z_q(\Lambda)$ and $\varphi.(B_q(\mathsf{K})) \subset B_q(\Lambda)$. In other words, cycles z_q of K go over into cycles $\varphi.z_q$ of Λ, and it follows from $z_q^1 \sim z_q^2$ that $\varphi.z_q^1 \sim \varphi.z_q^2$. Therefore, in every dimension q, $\varphi.$ induces a homomorphism of the homology groups, which we denote by φ_*,

$$\varphi_*\colon H_q(\mathsf{K}) \to H_q(\Lambda).$$

A simplicial mapping $\varphi\colon \mathsf{K} \to \Lambda$ therefore induces in all dimensions q a homomorphism φ_* of the corresponding homology groups.

§ 9. Examples and Applications

1. *The Zero-dimensional Groups* $H_0(\mathsf{K})$

9.1 Definition: If $c_0 = \sum g_i s_0^i$ $(i = 1, \ldots, \alpha_0)$ is a zero-dimensional chain of the complex K, then $\sum g_i = \text{In } c_0$ is called the (*Kronecker*) *index of the chain* c_0.

Since the boundary of a segment $\partial\langle \mathfrak{p}_0\mathfrak{p}_1\rangle = \langle \mathfrak{p}_1\rangle - \langle \mathfrak{p}_0\rangle$ always has the index 0, we have in general that $\text{In}(\partial c_1) = 0$ holds.— If K is connected, then, conversely, $\text{In } c_0 = 0$ implies $c_0 \sim 0$. Namely, since a fixed vertex $\mathfrak{p}_0$ can be joined to every other vertex $\mathfrak{p}$ by means of a polygonal arc, we have that $\langle \mathfrak{p}\rangle \sim \langle \mathfrak{p}_0\rangle$. For an arbitrary zero-dimensional chain $c_0 = \sum g_i s_0{}^i$ we then have $c_0 \sim (\text{In } c_0)\,\langle \mathfrak{p}_0\rangle$. Since $\langle \mathfrak{p}_0\rangle$ is not homologous to zero, c_0 is a boundary if and only if $\text{In } c_0 = 0$. The zero-dimensional homology classes therefore correspond in a one-to-one fashion to the values of $\text{In } c_0$, and we can state the following theorem.

9.2 Theorem: *The zero-dimensional homology group* $H_0(\mathsf{K})$ *of a connected complex* K *is isomorphic to* Z, *the cyclic group of the integers. Two zero-dimensional chains are homologous if and only if they have the same index.*

More generally, using an analogous line of reasoning one easily sees the following result.

9.3 Theorem: *If the complex* K *decomposes into* k *components, then* $H_0(\mathsf{K})$ *is the sum of* k *cyclic groups* Z. *We have that* $\wp_0(\mathsf{K}) = k$.

2. *Cone Formula and the Homology Groups of the Cone*

Let K be a simplicial complex in a realization in Euclidean N-space R^N. Take in $R^{N+1} \supset R^N$ a new vertex $\mathfrak{c}$ exterior to R^N and project all simplexes of K from $\mathfrak{c}$. There results a new simplicial complex, the *cone* over K as a base complex with $\mathfrak{c}$ as apex, which we denote briefly by $\mathfrak{c}\cdot\mathsf{K} = \mathsf{K}_1$. The cone K_1 consists of the following simplexes: First the 0-simplex $\mathfrak{c}$, second the simplexes of K, third the $(q+1)$-simplexes of the form $\mathfrak{c}\cdot\sigma^q$, which correspond in one-to-one fashion with the simplexes σ^q of K. Also to every chain c_q of K there is determined in an obvious way a $(q+1)$-dimensional chain

$\mathfrak{c} \cdot c_q$ of K_1. Then the following *cone formulas* hold:

$$(9.4) \qquad \begin{cases} \partial(\mathfrak{c} \cdot c_q) = c_q - \mathfrak{c} \cdot \partial c_q \text{ for } q \geqq 1, \\ \partial(\mathfrak{c} \cdot c_0) = c_0 - (\text{In } c_0) \langle \mathfrak{c} \rangle. \end{cases}$$

In the case $c_q = \langle \mathfrak{p}_0 \ldots \mathfrak{p}_q \rangle$, the proof is obtained directly by computation, namely we have that $\partial(\mathfrak{c} \cdot \langle \mathfrak{p}_0 \ldots \mathfrak{p}_q \rangle) = \langle \mathfrak{p}_0 \ldots \mathfrak{p}_q \rangle - \mathfrak{c} \cdot \partial \langle \mathfrak{p}_0 \ldots \mathfrak{p}_q \rangle$. The general case follows by means of a linear extension.

In order to compute the homology groups of a cone $\mathsf{K}_1 = \mathfrak{c} \cdot \mathsf{K}$, we again form a cone over K_1, namely $\mathsf{K}_2 = \mathfrak{t} \cdot \mathsf{K}_1$ with apex $\mathfrak{t}$, and define a simplicial mapping φ: $\mathsf{K}_2 \rightarrow \mathsf{K}_1$ by $\varphi(\mathfrak{t}) = \mathfrak{c}$ and $\varphi(\mathfrak{p}) = \mathfrak{p}$ for all vertices $\mathfrak{p} \in \mathsf{K}_1$. Then for every cycle z_q of K_1, considered as a chain of K_2, we have $\varphi.z_q = z_q$. Furthermore, for $q \geqq 1$, according to the cone formulas, we have that $\partial(\mathfrak{t} \cdot z_q) = z_q - \mathfrak{t} \cdot \partial z_q = z_q$. Upon application of $\varphi.$, there results $\varphi.[\partial(\mathfrak{t} \cdot z_q)] = \varphi.z_q = z_q$; therefore, $z_q = \partial(\varphi.(\mathfrak{t} \cdot z_q))$. Therefore, z_q is a boundary in K_1, i.e., every q-dimensional cycle is homologous to zero. In combination with Theorem 9.2, this yields the following theorem.

9.5 Theorem: *The homology groups of a cone* $\mathsf{K}_1 = \mathfrak{c} \cdot \mathsf{K}$ *are computed as follows:* $H_0(\mathsf{K}_1) \approx Z$, $H_q(\mathsf{K}_1) \approx 0$ *for* $q \geqq 1$.

3. *Homology Groups of an Element and of Spheres*

The complex Hu $\sigma^n = \mathsf{T}^n$ (see Volume I, § 30) of an n-dimensional simplex σ^n and all of its sides, which is a triangulation of the n-dimensional element $|\mathsf{T}|^n = T^n$, can obviously be considered as a cone with one vertex as apex and the opposite $(n-1)$-dimensional complex T^{n-1} as basic complex. Therefore the last theorem holds for T^n.—The same holds for the normal subdivision $\mathsf{T}^{n\prime}$ of T^n, which is clearly isomorphic to a cone whose apex corresponds to the center of gravity of σ^n.

The homology groups of the complex $\sum^n$ of all proper faces of an $(n+1)$-dimensional simplex σ^{n+1} with $n \geqq 1$ (see Volume I, § 30), which is a triangulation of the n-dimensional sphere $S^n = |\sum^n|$, can also be easily computed, if one considers $\sum^n$ as a subcomplex of T^{n+1}. Let $z_q \in Z_q(\sum^n)$ and $q \geqq 1$. According to the preceding theorem $z_q = \partial c_{q+1}$ in T^{n+1}. If $q < n$, then the chain c_{q+1} must lie in $\sum^n$; therefore $z_q \sim 0$ also holds in $\sum^n$. If $q = n$, $z_n = \partial c_{n+1}$, then $c_{n+1} = g s_{n+1}$ with integer g, where s_{n+1} is the elementary chain corresponding to σ^{n+1}, since there are no other $(n + 1)$-chains in T^{n+1}. Therefore $z_n = g \partial s_{n+1}$. Now ∂s_{n+1} is surely a cycle of $\sum^n$ and not a boundary of $\sum^n$ since absolutely no simplexes of the dimension $n + 1$ occur in $\sum^n$. Every cycle z_n is thus a multiple of the cycle ∂s_{n+1}. We thus see that $H_n(\sum^n)$ is infinite cyclic. We summarize in the next theorem.

9.6 Theorem: *The triangulations of the n-dimensional element* T^n *and* $\mathsf{T}^{n\prime}$ *have the homology groups* $H_0 \approx \mathsf{Z}$ *and* $H_q \approx 0$ *for* $q \neq 0$. *The triangulation* $\sum^n$ *of the n-dimensional sphere* $(n \geqq 1)$ *has the homology groups* $H_0 \approx H_n \approx Z$ *and* $H_q \approx 0$ *for* $q \neq 0, n$.

4. *Pseudomanifolds*

9.7 Definition: By an *n-dimensional pseudomanifold* $\mathsf{M} = \mathsf{M}^n$ we understand a subcomplex (Definition 3.3) M of a simplicial complex K with the following properties:

(a) M is homogeneously n-dimensional, i.e., every simplex of M is the face of an n-dimensional simplex of M.

(b) M is unramified, i.e., every $(n - 1)$-simplex of M is incident with exactly two n-simplexes.

(c) M is strongly connected, i.e., every two n-simplexes of M can be joined by an alternating sequence of n-dimensional and $(n - 1)$-dimensional incident simplexes.

If $\mathsf{M} = \mathrm{Hu}\mathsf{M}$, i.e., M is itself a simplicial complex, then M is called a *closed* n-dimensional pseudomanifold.

A pseudomanifold is, therefore, a subset of a simplicial complex K, the embedding complex, and, in general, is not itself a complex. The pseudomanifold M is itself a complex if and only if $\mathsf{M} = \mathrm{Hu}\,\mathsf{M}$; in this case, we can regard M itself as the embedding complex.

Examples: (1) For $n = 0$, a pseudomanifold M^n is simply a single point.—(2) In the case $n = 1$, the simple closed polygonal curves are the only closed pseudomanifolds.—(3) In the case $n = 2$, the closed surfaces, the 2-spheres, the torus, the two-holed tori, and so forth, yield pseudomanifolds upon triangulation.—(4) Another example of a closed pseudomanifold M^2 is obtained if we rotate a circle about one of its tangents and then triangulate the surface which arises. One should convince himself that the conditions (a), (b), (c) are also satisfied in the midpoint of this surface.—The following are examples of non-closed pseudomanifolds M^n: (5) A single simplex σ^n of an arbitrary complex.—(6) The star of an arbitrary simplex σ^q in an arbitrary simplicial decomposition of the n-sphere.

Let M^n be a pseudomanifold. It can happen that the n-simplexes of M^n can be so oriented that every $(n - 1)$-simplex appears with opposite sign in the boundary of the two n-simplexes incident with it (see (b)) with opposite sign. The n-simplexes of M^n are then said to be *coherently oriented*; M^n is said to be *orientable*. There are obviously exactly two distinct coherent orientations of an orientable M^n, which arise from one another by reversal of the orientation of all the n-simplexes. If one of the two orientations is chosen fixed, then we say that M^n is *oriented*.—If a coherent orientation of M^n is not possible, then M^n is said to be *non-orientable*. An example of a non-orientable pseudomanifold is formed by a Möbius band in any simplicial decomposition, in which the boundary curve is not included.

We now investigate the homology groups of closed pseudomanifolds. (For the time being, the homology groups for non-closed pseudomanifolds are not defined, only later will we assign homology groups to arbitrary subcomplexes.) Therefore, let M^n be a closed pseudomanifold. Moreover, let M^n be orientable and have a fixed orientation. The totality of n-simplexes in coherent orientation form a cycle μ_n. On the other hand, if an arbitrary n-dimensional cycle z_n of M^n contains any n-simplex k times, then it must also contain every neighboring n-simplex and thus every n-simplex of M^n k times. Therefore, we have $z_n = k\mu_n$ $(k = 0, \pm 1, \ldots)$, i.e. every n-dimensional cycle of M^n is a multiple of μ_n. This cycle is called the *basic cycle* μ_n of M^n. (The notation μ_n, which deviates from our usual notation, is to emphasize the fact that μ_n corresponds to M^n.) M^n, together with the orientation, determines the basic cycle uniquely; in accordance with the two orientation possibilities, M^n determines the basic cycle only up to sign.—Since $H_n(\mathsf{Z}^n) = \mathsf{Z}_n(\mathsf{M}^n)$, we have $H_n(\mathsf{M}^n)$ is already determined, $H_n(\mathsf{M}^n) \approx \mathsf{Z}$.—If the pseudomanifold is not orientable, then we cannot, of course, give an n-dimensional cycle of M^n, inasmuch as such a cycle would yield a coherent orientation of M^n. Thus, in this case, we have $H_n(\mathsf{M}^n) \approx 0$. —For the zero-dimensional homology group, whether M^n is orientable or not, $H_0(\mathsf{M}^n) \approx \mathsf{Z}$ because M^n is a connected complex. We note the following theorem.

9.8 Theorem: *In an orientable closed n-dimensional pseudomanifold* M^n *every n-dimensional cycle is an integral multiple of the basic cycle* μ_n *of* M^n, *which is determined up to sign by* M_n *and which, for a fixed orientation of* M^n, *is uniquely determined. We have that* $H_0(\mathsf{M}^n) \approx H_n(\mathsf{M}^n) \approx \mathsf{Z}$.

The pseudomanifolds given above as examples (1)–(6) are all orientable. In order to also give an example of a non-orientable closed pseudomanifold, we consider first the *Möbius band* shown in the adjacent figure (see Fig. 3). It arises by cutting an ordinary closed band and then

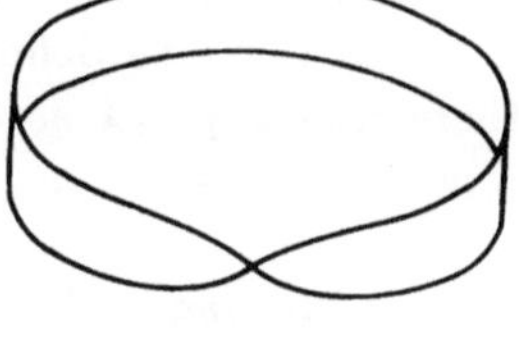

Fig. 3

putting it together again twisted. The Möbius band is bound by a single closed curve.—We now add to this closed curve an open circular disk. Indeed, this is impossible in Euclidean R^3 without interpenetration, but it can be done in an R^4 in which R^3 in embedded, namely by means of a projection of this closed curve from a point lying in R^4 but not in R^3.

Thus, there arises in R^4 a surface, the *closed Möbius band* B_1. The surface B_1 can easily be decomposed simplicially and yields then a 2-dimensional non-orientable pseudomanifold B_1. The Betti numbers of B_1 are $\wp_0 = 1$ (Theorem 9.2), $\wp_1 = \wp_2 = 0$; there is a 1-dimensional torsion coefficient of 2, i.e., $H_1(\mathsf{B}_1)$ is cyclic of order 2. Here is an example of a 1-dimensional cycle z_1 which is not a boundary but whose doubling $2z_1$ is a boundary.—Moreover, the closed Möbius band B_1 is homeomorphic to the projective plane.

5. *The Mapping Degree* γ

Let f: $\mathsf{M}^n \to \mathsf{N}^n$ be a simplicial mapping of an oriented closed n-dimensional pseudomanifold into another pseudomanifold of this type. If μ_n and ν_n are the basic cycles of M^n and N^n, respectively, then the mapping f effects a relation of the form $f.\mu_n = \gamma\nu_n$ with $\gamma = \gamma(f)$ an integer. Similarly, making use of f_*: $H_n(\mathsf{M}^n) \to H_n(\mathsf{N}^n)$, we can write $f_*\mu_n = \gamma\nu_n$. The integer γ is called the *mapping degree*, or briefly the *degree*, of f and will turn out to be an important piece of information for determining f.

Examples: (1) For $n > 0$, a constant mapping of M^n into N^n has the degree 0. The identity mapping has the degree 1.—(2) The reflection of the 2-sphere in its equator has the degree -1.—(3) Let M' and N' be simplicially decomposed circles whose points are determined in the usual way by means of the argument angles φ and ψ as coordinates. For every $k = 0, \pm 1, \pm 2, \ldots$, let a mapping $\psi = f_k(\varphi) = k\varphi$ of $|\mathsf{M}'|$ into $|\mathsf{N}'|$ be defined. Under a simplicial approximation of f_k there results $\gamma = k$ as the mapping degree.

The following theorem is obvious.

9.9 Theorem: *If* f: $\mathsf{M}^n \to \mathsf{N}^n$ *and* g: $\mathsf{N}^n \to \Lambda^n$ *are mappings of closed oriented n-dimensional pseudomanifolds with the mapping degrees* γ *and* δ, *then* gof: $\mathsf{M}^n \to \Lambda^n$ *has the mapping degree* $\delta\gamma$.

6. *Some Special Polyhedra*

As examples, we give several simple 1-, 2- and 3-dimensional polyhedra $R = |\mathsf{K}|$ and their Betti numbers. More specifically stated, we give the Betti numbers of the simplicial decomposition which, as we shall see later, correspond to the polyhedron R as an invariant and, therefore, are the same for all simplicial decompositions K of the polyhedron R. We combine them into the *Poincaré polynomial* $\wp(x) = \sum \wp_q(\mathsf{K})x^q$, summed over all dimensions q. Here x signifies an indeterminate to which no geometric interpretation is attached. The reader can verify for himself the given values of the $\wp_q$ as the "number of linearly independent closed, but not bounding, chains" by means of a suitable simplicial decomposition K of R. An exact computation of these values is, at times, very wearisome using the methods we developed up to this point. In the next chapters, we will get acquainted with simple and effective methods of computation. They stem from the determination of the rank of matrices and will be especially simple in application to the cell decomposition of §§18, 19 (compare primarily the last note).—(a) Let R be a single point: $\wp(x) = 1$.—(b) Let R be a segment: $\wp(x) = 1$.—(c) Let R be a circle: $\wp(x) = 1 + x$.—(d) Two circles with a common point: $\wp(x) = 1 + 2x$.—(e) Circular disk: $\wp(x) = 1$.—(f) Circular disk with a smaller circular disk cut out (plane annulus): $\wp(x) = 1 + x$.—(g) Circular disk with two smaller circular disks cut out: $\wp(x) = 1 + 2x$.—(h) 2-sphere: $\wp(x) = 1 + x^2$.—(i) 2-sphere with two circular disks cut out (2-sphere with two holes): $\wp(x) = 1 + x$.—(j) Torus: $\wp(x) = 1 + 2x + x^2$.—(k) Two-holed torus (see adjacent figure, Fig. 4): $\wp(x) = 1 + 4x + x^2$.—(l) Solid sphere: $\wp(x) = 1$.—

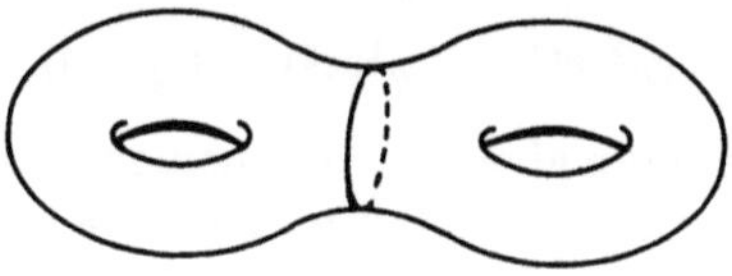

Fig. 4

(m) Solid sphere with a smaller solid sphere cut out (hollow sphere): $\wp(x) = 1 + x^2$.—(n) Solid torus: $\wp(x) = 1 + x$.—(o) Solid two-holed torus: $\wp(x) = 1 + 2x$.—(p) Solid sphere with solid torus cut out: $\wp(x) = 1 + x + x^2$.—Note for instance that the polyhedra (f) and (i) are homeomorphic and that furthermore for example the polyhedron (c) is a deformation retract of the polyhedron (f) and of the polyhedron (n). We can construct further examples by generalization to higher dimensions. Another procedure for constructing new polyhedra is the formation of the topological product; in this procedure, the Poincaré polynomials are multiplied together (stated here without proof).—All these examples yield, of course, only a small portion of the unlimited set of polyhedra.

§ 10. Cohomology Groups

Again, let K be a fixed n-dimensional simplicial complex. First, we give two distinct but coextensive definitions of a cochain of K. 1. *Definition:* A *q-dimensional cochain* c^q is a function which assigns to each positively oriented simplex s_q^i ($i = 1, \ldots, \alpha_q$) of K an integer $(s_q^i, c^q) = g^i$.—Among the cochains there appear, in particular, the elementary cochains s_j^q ($j = 1, \ldots, \alpha_q$) which assign the value 1 to the simplex s_q^j and the value 0 to all other q-simplexes: $(s_q^i, s_j^q) = \delta_{ij}$, where the Kronecker symbol δ_{ij} denotes 1 if $i = j$ and 0 if $i \neq j$.

By a linear extension, each cochain c^q assigns to each chain $c_q = \sum h_i s_q^i$ an integer, namely by means of the definition

$$(c_q, c^q) = (\sum h_i s_q^i, c^q) = \sum h_i (s_q^i, c^q).$$

In particular, we have that $(-s_q^i, c^q) = -(s_q^i, c^q)$. We also see that

$$(10.1) \qquad (c_q^1 + c_q^2, c^q) = (c_q^1, c^q) + (c_q^2, c^q).$$

The cochain c^q therefore gives rise to a homomorphism of the group C_q into the cyclic group Z of integers. Thus, we arrive at the second definition of a cochain. 2. *Definition:* A *cochain* c^q is a homomorphism $c^q\colon C_q \to \mathsf{Z}$ of C_q into the group of integers Z which assigns to each chain $c_q \in C_q$ an integer (c_q, c^q).—If such a homomorphism c^q is given and if we designate the values (s_q^i, c^q) by g^i, then c^q yields a function in the sense of the first definition. This function, as has just been described, again leads back to the homomorphism c^q. Thus, the definitions are equivalent. We will use one or the other according to expediency. We call (c_q, c^q) the *Kronecker product* of the chain c_q with the cochain c^q.

We can add and subtract the functions c^q and hence the homomorphisms c^q in the usual manner, by letting $(c_q, -c^q) = -(c_q, c^q)$ and

$$(c_q, c_1{}^q + c_2{}^q) = (c_q, c_1{}^q) + (c_q, c_2{}^q). \tag{10.2}$$

Thus, the cochains c^q form a group C^q. For a given c^q, if we consider the values $(s_q{}^i, c^q) = g^i$, we recognize that each c^q possesses a unique representation in terms of the elementary cochains $s_i{}^q$, $c^q = \sum g^i s_i{}^q$, which is entirely analogous to the representation of the chains $c_q = \sum h_i s_q{}^i$ in terms of a chain basis $s_q{}^i$. Therefore, C^q is a free abelain group and a module. The elementary cochains $s_i{}^q$ form a basis for C^q.

The relation between C_q and C^q can be also represented in symmetric form. The Kronecker product (c_q, c^q) yields a *pairing* of C_q and C^q which assigns to each pair c_q, c^q an integer. In this connection, the rules (10.1) and (10.2) are valid. We say that the pairing by means of the Kronecker product is bilinear.

We now come to the *coboundary operator* δ which assigns to each q-dimensional cochain c^q a $(q + 1)$-dimensional cochain $\delta c^q = c^{q+1}$. We define δc^q to be the $(q + 1)$-dimensional cochain such that

$$(c_{q+1}, \delta c^q) = (\partial c_{q+1}, c^q) \qquad \text{(Stokes's Formula)} \tag{10.3}$$

for every chain c_{q+1}. We see that this is really a homomorphism if we replace c_{q+1} by $c^1_{q+1} + c^2_{q+1}$ and observe the effect on the right side. For a better understanding of the Stokes's formula, consider the special case of an elementary chain s_{q+1}:

$$(\langle \mathfrak{p}_0 \ldots \mathfrak{p}_{q+1} \rangle, \delta c^q) = \sum_{i=0}^{q+1} (-1)^i (\langle \mathfrak{p}_0 \ldots \hat{\mathfrak{p}}_i \ldots \mathfrak{p}_{q+1} \rangle, c^q). \tag{10.4}$$

This asserts that *the function δc^q assumes on a simplex s_{q+1} that value which is equal to the sum of the values of c^q on the q-dimensional* *boundary simplexes (in the induced orientation) of* s_{q+1}. This corresponds to a large extent to the boundary formation ∂c_q, which can be described as follows: *The chain* ∂c_q *has on a simplex* s_{q-1} *a coefficient which is equal to the sum of the coefficients of* c_q *on the q-dimensional simplexes (in the proper orientation) bounded by* s_{q-1}.—The homomorphisms δ taken together form a sequence

$$0 \overset{\delta}{\leftarrow} C^n \overset{\delta}{\leftarrow} C^{n-1} \overset{\delta}{\leftarrow} \cdots \overset{\delta}{\leftarrow} C^q \overset{\delta}{\leftarrow} \cdots \overset{\delta}{\leftarrow} C^0 \overset{\delta}{\leftarrow} 0.$$

We now prove that $\delta\delta = \delta^2 = 0$. This follows immediately from the equation $\partial^2 = 0$ by applying the Stokes's formula twice,

$$(c_{q+2}, \delta^2 c^q) = (\partial c_{q+2}, \delta c^q) = (\partial^2 c_{q+2}, c^q) = 0.$$

Cochains z^q with $\delta z^q = 0$ are called *cocycles*. The totality of q-dimensional cocycles of K form a group Z^q. Cochains c^q which are representable in the form $c^q = \delta c^{q-1}$ are called *coboundaries*. The totality of q-dimensional coboundaries form a group B^q. Obviously, $B^0 = 0$ and, if K is n-dimensional, then $Z^n = C^n$. Every coboundary $c^q = \delta c^{q-1}$ is also a cocycle because $\delta c^q = \delta^2 c^{q-1} = 0$. Therefore, B^q is a subgroup of Z^q. The factor group $Z^q/B^q = H^q$ is called the *q-dimensional cohomology group* $H^q = H^q(\mathsf{K})$ *of* K. Assertions, corresponding to those made in the preceding section about the homology groups, can be made about the cohomology groups. Two cocycles from the same cohomology class are said to be cohomologous, or simply, homologous.—For the extreme dimensions $q = 0$ and $q = n$, we have:

(10.5) $$H^0 = Z^0, H^n = C^n/B^n.$$

If, for instance, K is a connected simplicial complex, then we have the following immediate analogy to Theorem 9.2: The 0-dimensional cochain c^0 which has the value $+1$ on all vertices of K is a cocycle and every other zero-dimensional cocycle of K is a multiple of this cocycle c^0. *For a connected simplicial complex* K, *we have that* $H^0(\mathsf{K}) = Z^0(\mathsf{K}) \approx \mathsf{Z}$.—Furthermore, if M^n is a connected closed orientable pseudomanifold, then analogous to Theorem 9.8, the following is valid. Let $s_i{}^n$ denote that cocycle which has the value

$+1$ on the n-simplex $s_n{}^i$ and the value 0 on the other n-simplexes of M^n. Because of the strong connectivity of M^n (condition (c) of Definition 9.7), every two cocycles of this sort are mutually homologous, and for that reason an arbitrary cocycle $\sum g^i s_i{}^n$, with integer g^i and summed over i, is homologous to $(\sum g^i)s_1{}^n$. Therefore, two n-dimensional cocycles are homologous if and only if the sum of their coefficients $\sum g^i$ is the same. This means that *for a connected closed orientable pseudomanifold* M^n, *we have that* $H^n(\mathsf{M}^n) = C^n/B^n \approx \mathsf{Z}$.

Before continuing, we point out a remarkable analogy of homology theory and cohomology theory with the theory of multidimensional integrals. This analogy contributes to the understanding of homology and cohomology and especially their mutual interrelationships. On the other hand, the theory of integration is illuminated by this analogy from the algebraic point of view. The zero-dimensional cochains c^0 from K represent a discontinuous analogue of the scalar functions in the space $R = |\,\mathsf{K}\,|$. There is no easy way to clarify this and it is inessential to the further development. In the same way, the 1-dimensional cochains c^1 form a discontinuous analogue of a vector field in R insofar as they assign a numerical value to each directed edge of K. The c^q correspond to the q-dimensional functions, or better yet, to the q-dimensional differential forms in R. If we conceive of the chains c_q of K as q-dimensional portions of space which are equipped with a certain multiplicity in each q-simplex, then the Kronecker product corresponds to the definition of sums of the definite integral; more specifically,

$$(c_q, c^q) = \sum_{i=1}^{\alpha} g_i\,(s_q{}^i, c^q) \qquad \text{and} \qquad \int_{c_q} c^q$$

correspond to one another. The Kronecker product, therefore, appears as a discontinuous analogue of the q-dimensional integral. The rules 10.1 and 10.2 correspond to the additive properties of the integral relative to the domain of integration and to the differential form respectively. The operator ∂ corresponds to the boundary formation in portions of space, and $\partial^2 = 0$ is illuminating for this reason. We must conceive of δc^q as a differential operation which assigns to a q-dimensional differential c^q the $(q + 1)$-dimensional differential δc^q. In the case of R^3, there corresponds to δc^0 the *gradient* operator, to δc^1 the *curl* operator, and to δc^2 the *divergence* operator. In this connection, $\delta^2 = 0$ corresponds to either the equation curl grad $= 0$ or to the equation div curl $= 0$. A one-dimensional cocycle z^1 with $\delta z^1 = 0$ represents a vortex-free vector field. Furthermore, the formulas

$$(c_{q+1}, \delta c^q) = (\partial c_{q+1}, c^q) \quad \text{and} \quad \int_{c_{q+1}} \delta c^q = \int_{\partial c_{q+1}} c^q$$

correspond to one another. The latter is the *Stokes's integral theorem.* Differential forms c^q with $\delta c^q = 0$ are called *exact* and those which admit a representation $c^q = \delta c^{q-1}$ are called *total.* The q-dimensional cohomology group $H^q(\mathsf{K})$, therefore, corresponds to the group of exact, but not total, differential forms. The homology group $H_q(\mathsf{K})$ corresponds to the group of the closed, but not bounding, domains of integration. We can develop this analogy further by taking real numbers instead of the integers as the basis for the coefficients of the chains and as values of cochains. The transition to the integral formulas is difficult and leads to the theory of harmonic integrals which naturally lies far outside the framework of this little volume.

Stokes's formula (10.3) has another important application. Suppose the cycle z_q represents one of the homology classes of $H_q(\mathsf{K})$. Then any cycle of the same class is of the form $z_q + \partial c_{q+1}$ for some chain c_{q+1}. If z^q represents a cohomology class, then any of the remaining cocycles are of the form $z^q + \delta c^{q-1}$ for some cochain c^{q-1}. Then

$$(z_q + \partial c_{q+1}, z^q + \delta c^{q-1}) = (z_q, z^q) + (z_q, \delta c^{q-1}) + \\ + (\partial c_{q+1}, z^q) + (\partial c_{q+1}, \delta c^{q-1}).$$

The last three summands vanish; for instance,

$$(z_q, \delta c^{q-1}) = (\partial z_q, c^{q-1}) = 0.$$

Therefore, the Kronecker product (z_q, z^q) does not depend on the particular choice of the chains z_q and cochains z^q in their respective homology classes and cohomology classes. Thus, the Kronecker products yields a scalar multiplication between the homology classes and cohomology classes which we shall also call the *Kronecker product*. We now formulate the next theorem.

10.6 Theorem: *The Kronecker product between cycles and cocycles of* K *yields a bilinear product, defined in terms of representatives, between the homology classes of* $H_q(\mathsf{K})$ *and the cohomology classes of* $H^q(\mathsf{K})$.

10.7 Theorem: *If* $(c_{q+1}, \delta c^q) = 0$ *for all* c^q, *then* c_{q+1} *is a cycle. If* $(\partial c_{q+1}, c^q) = 0$ *for all* c_{q+1}, *then* c^q *is a cocycle.*

Proof: If the equation $0 = (c_{q+1}, \delta s_i^q) = (\partial c_{q+1}, s_i^q)$ holds for all elementary cochains s_i^q, then this signifies that the coefficient of ∂c_{q+1} of each simplex s_q^i is zero, i.e., that $\partial c_{q+1} = 0$.—The second part of the theorem is proved similarly.

In conclusion, it is shown that as in the case of homology groups a simplicial mapping φ of a simplicial complex K into a simplicial complex Λ induces a homomorphism on the cohomology groups of K and Λ as well as on the homology groups. But in the case of cohomology groups, the direction is from Λ to K. Namely, let $d^q \in C^q(\Lambda)$. We make correspond to d^q a cochain $c^q = \varphi{\cdot}d^q \in C^q(\mathsf{K})$, defined by prescribing the value of c^q on an arbitrary chain $c_q \in C_q(\mathsf{K})$; moreover, the procedure is entirely analogous to that in the definition of δc^q above:

$$(c_q, \varphi{\cdot}d^q) = (\varphi.c_q, d^q). \tag{10.8}$$

If one replaces c_q by $c_q{}^1 + c_q{}^2$ and observes the effect on the right side, it is clear that a homomorphism c^q of $C_q(\mathsf{K})$ into Z is thus defined. To the sum $d_1{}^q + d_2{}^q$ of two elements of $C^q(\Lambda)$ there corresponds, under $\varphi{\cdot}$, the sum $\varphi{\cdot}d_1{}^q + \varphi{\cdot}d_2{}^q$ since

$$(c_q, \varphi{\cdot}(d_1{}^q + d_2{}^q)) = (\varphi.c_q, d_1{}^q + d_2{}^q) = (\varphi.c_q, d_1{}^q) + \\ + (\varphi.c_q, d_2{}^q) = (c_q, \varphi{\cdot}d_1{}^q) + (c_q, \varphi{\cdot}d_2{}^q).$$

For this reason, $\varphi^{\cdot}$ is indeed a homomorphism, $\varphi^{\cdot}: C^q(\Lambda) \to C^q(\mathsf{K})$.

We compare further the two mappings $\varphi.$ and $\varphi^{\cdot}$. We have that $\varphi.s_q$ is the image simplex of s_q under φ. The $\varphi^{\cdot}$-image $\varphi^{\cdot}t^q$ of an elementary cochain t^q of Λ has the value 1 for those K-simplexes which go over into t_q under $\varphi.$.—The mapping $\varphi^{\cdot}$ preserves coboundaries, i.e., $\delta\varphi^{\cdot} = \varphi^{\cdot}\delta$ is valid. This results if we apply the Stokes's formula and the fact that $\varphi.$ preserves boundaries:

$$\begin{aligned}(c_{q+1}, \varphi^{\cdot}\delta d^q) &= (\varphi.c_{q+1}, \delta d^q) = (\partial\varphi.c_{q+1}, d^q) = (\varphi.\partial c_{q+1}, d^q)\\ &= (\partial c_{q+1}, \varphi^{\cdot}d^q) = (c_{q+1}, \delta\varphi^{\cdot}d^q).\end{aligned}$$

Here, analogous to the situation in homology theory, there results a commutative diagram:

$$\begin{array}{ccccccccccc} 0 & \overset{\delta}{\leftarrow} & C^n(\mathsf{K}) & \overset{\delta}{\leftarrow} & C^{n-1}(\mathsf{K}) & \overset{\delta}{\leftarrow} & \cdots & \overset{\delta}{\leftarrow} & C^0(\mathsf{K}) & \overset{\delta}{\leftarrow} & 0 \\ \varphi.\uparrow & & \varphi^{\cdot}\uparrow & & \varphi.\uparrow & & & & \varphi.\uparrow & & \varphi^{\cdot}\uparrow \\ 0 & \overset{\delta}{\leftarrow} & C^n(\Lambda) & \overset{\delta}{\leftarrow} & C^{n-1}(\Lambda) & \overset{\delta}{\leftarrow} & \cdots & \overset{\delta}{\leftarrow} & C^0(\Lambda) & \overset{\delta}{\leftarrow} & 0. \end{array}$$

In a manner corresponding to $\varphi.$ and the homology groups, it now follows that $\varphi^{\cdot}(Z^q(\Lambda)) \subset Z^q(\mathsf{K})$, $\varphi^{\cdot}(B^q(\Lambda)) \subset B^q(\mathsf{K})$. Thus, $\varphi^{\cdot}$ induces a homomorphism for the cohomology groups,

$$\varphi^*: H^q(\Lambda) \to H^q(\mathsf{K}).$$

Therefore, a simplicial mapping φ: $\mathsf{K} \to \Lambda$ induces a homomorphism φ^* of the cohomology groups in the reversed direction.

§ 11. Homotopic Mappings

11.1 Theorem: *Let φ, ψ: $\mathsf{K} \to \Lambda$ be two simplicial mappings of the same neighborhood class. Then, for each q, there exist homomorphisms D_q: $C_q(\mathsf{K}) \to C_{q+1}(\Lambda)$ with the property $\partial D_q + D_{q-1}\partial = \psi. - \varphi.$, or briefly, $\partial D + D\partial = \psi. - \varphi.$.*

The meaning of this homomorphism D, the *homotopy operator*, and the given operator equation for $\varphi.$ and $\psi.$ is first motivated by the sketch on Fig. 5. Let $\varphi.\mathfrak{c}_1$ and $\psi.\mathfrak{c}_1$ be the two mutually homotopic images of a chain $\mathfrak{c}_1$ of K into Λ. Under the homotopy, $\varphi.\mathfrak{c}_1$ is carried over continuously into $\psi.\mathfrak{c}_1$; think of this carrying over say as a continuous movement from $\varphi.\mathfrak{c}_1$ to $\psi.\mathfrak{c}_1$. The region hatched over in this connection represents $D\mathfrak{c}_1$. The boundary points of $\varphi.\mathfrak{c}_1$ represent $\partial\varphi.\mathfrak{c}_1$ with the sign given in the figure. Under the deformation, they pass over the segments $D\partial\mathfrak{c}_1$ which have the given

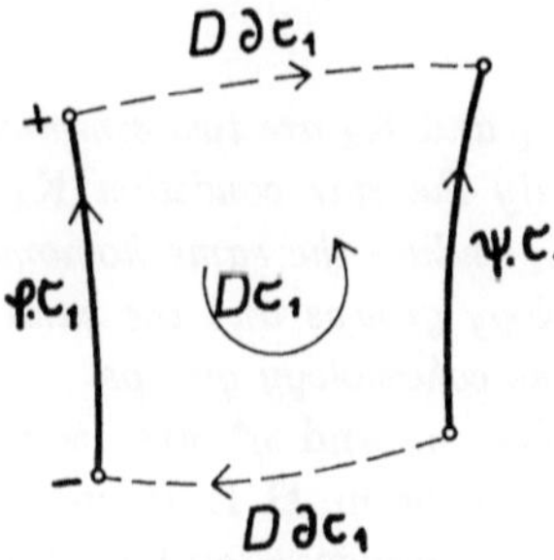

Fig. 5

orientations. Then the sketch shows that $\partial D\mathfrak{c}_1 = \psi.\mathfrak{c}_1 - \varphi.\mathfrak{c}_1 - D\partial\mathfrak{c}_1$ is valid, which corresponds to the operator equation of the theorem.—We postpone the proof of the theorem and deduce a corollary first.

11.2 Theorem: *If the simplicial mappings φ, ψ: $\mathsf{K} \to \mathsf{\Lambda}$ belong to the same neighborhood class, or, more generally, if φ and ψ are interconnected by a homotopy operator D as in the last theorem, then φ and ψ induce the same homomorphism $\varphi_* = \psi_*$.*

Proof: Namely, if $z_q \in Z_q(\mathsf{K})$ is a q-dimensional cycle of K, then there results the operator equation $\psi.z_q = \varphi.z_q + \partial Dz_q + D\partial z_q$. The last summand vanishes since $\partial z_q = 0$; the next to the last summand is a boundary. Therefore, $\psi.z_q \sim \varphi.z_q$ for every cycle z_q, which completes the proof of the theorem.

The Theorems 11.2 and 6.3 imply the part of the following theorem concerning homology groups. Regarding the cohomology groups, compare the remark at the conclusion of this section.

11.3 Theorem: *Every two simplicial approximations φ, ψ: $\mathsf{K} \to \Lambda$ of a continuous mapping f: $|\mathsf{K}| \to |\Lambda|$ induce the same homomorphism of the homology groups, $\varphi_* = \psi_*$, and the same homomorphism of the cohomology groups, $\varphi^* = \psi^*$.*

For later applications, we formulate the following more particular theorem.

11.4 Theorem: *If K_1 and K_2 are two simplicial subdivisions of a polyhedron which satisfy the star condition $\mathsf{K}_1 < \mathsf{K}_2$, then all displacements η: $\mathsf{K}_1 \to \mathsf{K}_2$ induce the same homomorphism η_*: $H_q(\mathsf{K}_1) \to H_q(\mathsf{K}_2)$ of the homology groups and the same homomorphism η^*: $H^q(\mathsf{K}_1) \leftarrow H^q(\mathsf{K}_2)$ of the cohomology groups.*

We will see later that η_* and η^* are isomorphisms.—Now we come to the proof of Theorem 11.1. It suffices to carry out the proof for neighboring mappings φ and ψ. For, if φ_1 and φ_2 are neighboring mappings and D_1 is a corresponding homotopy operator with $\partial D_1 + D_1\partial = \varphi_2. - \varphi_1.$, and if φ_2 and φ_3 are neighboring mappings with the homotopy operator D_2 and $\partial D_2 + D_2\partial = \varphi_3. - \varphi_2.$, then it follows, by addition of the two equations, that $\partial(D_1 + D_2) + (D_1 + D_2)\partial = \varphi_3. - \varphi_1.$, where $(D_1 + D_2)$: $C_q(\mathsf{K}) \to C_{q+1}(\Lambda)$ where $(D_1 + D_2)c_q = D_1c_q + D_2c_q$ for all $c_q \in C_q(\mathsf{K})$ for all q. Therefore, there exist for φ_1 and φ_3 a homotopy operator, namely $D_1 + D_2$. This procedure carries over directly to finite sequences of neighboring mappings.

Thus, let φ and ψ be neighboring mappings. If $s_q = \langle \mathfrak{p}_0 \dots \mathfrak{p}_q \rangle$ is a simplex of K, then the vertices $\varphi(\mathfrak{p}_i) = \mathfrak{q}_i$ and $\psi(\mathfrak{p}_i) = \mathfrak{r}_i$ ($i = 0, \dots, q$) lie on a simplex of Λ. There are formally $2(q + 1)$ distinct vertices, but they may coincide in part. We now define the homotopy operator D by means of the following equation

$$Ds_q = \sum_{i=0}^{q} (-1)^i \langle \mathfrak{q}_0 \dots \mathfrak{q}_i\mathfrak{r}_i \dots \mathfrak{r}_q \rangle$$

and by linear extension to the entire group $C_q(\mathsf{K})$. If several vertices coincide in $\langle \mathfrak{q}_0 \dots \mathfrak{r}_q \rangle$, replace the expression by zero. Since we saw in §8 that such expressions yield boundary zero under formal boundary formation, we will not change the following computation. We must show that D satisfies the operator equation of Theorem 11.1. It suffices to check this equation for s_q. We now have that

$$\partial D s_q = \sum_{i=0}^{q} (-1)^i \partial \langle \mathfrak{q}_0 \ldots \mathfrak{q}_i \mathfrak{r}_i \ldots \mathfrak{r}_q \rangle,$$

$$(11.5) \quad \partial D s_q = \sum_{i=0}^{q} (-1)^i \Bigg[\sum_{j=0}^{i-1} (-1)^j \langle \mathfrak{q}_0 \ldots \hat{\mathfrak{q}}_j \ldots \mathfrak{q}_i \mathfrak{r}_i \ldots \mathfrak{r}_q \rangle$$

$$+ (-1)^i \langle \mathfrak{q}_0 \ldots \mathfrak{q}_{i-1} \mathfrak{r}_i \ldots \mathfrak{r}_q \rangle + (-1)^{i+1} \langle \mathfrak{q}_0 \ldots \mathfrak{q}_i \mathfrak{r}_{i+1} \ldots \mathfrak{r}_q \rangle$$

$$+ \sum_{j=i+1}^{q} (-1)^{j+1} \langle \mathfrak{q}_0 \ldots \mathfrak{q}_i \mathfrak{r}_i \ldots \hat{\mathfrak{r}}_j \ldots \mathfrak{r}_q \rangle \Bigg].$$

The members of the middle row, summed over $i = 0, \ldots, q$, cancel each other out in pairs, leaving only $\langle \mathfrak{r}_0 \ldots \mathfrak{r}_q \rangle - \langle \mathfrak{q}_0 \ldots \mathfrak{q}_q \rangle$. On the other hand, we have

$$\partial s_q = \sum_{j=0}^{q} (-1)^j \langle \mathfrak{p}_0 \ldots \hat{\mathfrak{p}}_j \ldots \mathfrak{p}_q \rangle, \quad \text{and}$$

$$D \partial s_q = \sum_{j=0}^{q} (-1)^j \Bigg[\sum_{i=0}^{-1} (-1)^i \langle \mathfrak{q}_0 \ldots \mathfrak{q}_i \mathfrak{r}_i \ldots \hat{\mathfrak{r}}_j \ldots \mathfrak{r}_q \rangle$$

$$+ \sum_{i=j+1}^{q} (-1)^{i-1} \langle \mathfrak{q}_0 \ldots \hat{\mathfrak{q}}_j \ldots \mathfrak{q}_i \mathfrak{r}_i \ldots \mathfrak{r}_q \rangle \Bigg].$$

The first sum, upon reversal of the sign, is identical with the last row of (11.5). Similarly, upon reversal of the sign, the second summand is identical with the first row of (11.5). This completes the proof of the theorem.

The considerations of this section on homology groups carry over meaningfully to cohomology groups as well. Since the process is parallel step by step, we need not carry it out here in detail but content ourselves with the assertion on the induced homomorphism of Theorem 11.3.

PART II

CHAIN COMPLEXES AND THEIR APPLICATIONS

Chapter 3

GENERAL THEORY

§ 12. Homology Groups of Chain Complexes

In the preceding chapter, algebraic constructions, namely homology groups, were made to correspond to polyhedra. In this and the next chapter, homology groups and their properties will be considered from the purely algebraic point of view. For this purpose, we construct a purely algebraic model, the chain complexes, in which the homology groups are defined anew and then investigated in more detail. This does not occur from independent algebraic interest. Rather, our algebraic considerations will be isolated and thereby made more understandable, that is, they will be placed independently in juxtaposition to the purely geometric portion. In this way, many repetitions of auxiliary algebraic considerations will also be avoided, and the common algebraic sources of various facts will be made clear.

12.1 Definition: A sequence $\mathfrak{C}$ of abelian groups C_q and homomorphisms $\partial: C_q \to C_{q-1}$,

$$0 \to C_n \to C_{n-1} \to \ldots \to C_q \to \ldots \to C_1 \to C_0 \to 0,$$

with the property that $\partial\partial = \partial^2 = 0$ is called an *n-dimensional chain complex*. The elements of C_q are called *q-dimensional chains* (or briefly *q-chains*), q is called the *dimension* of C_q, and ∂ is called the *boundary operator*.

The 0-groups at the beginning and end of the sequence are adjoined out of purely formal considerations; for convenience, we set $C_q = 0$

for $q < 0$ and for $q > n$.—A chain $c_q \in C_q$ with $\partial c_q = 0$ is called a *q-dimensional cycle*; the subgroup of all q-dimensional cycles, that is, the kernel of the homomorphism $\partial: C_q \to C_{q-1}$, is called the *group Z_q of the q-dimensional cycles* of $\mathfrak{C}$. A chain c_q of C_q, for which there exists a chain $c_{q+1} \in C_{q+1}$ with $\partial c_{q+1} = c_q$, is called a *bounding chain* or a *boundary chain*. The subgroup of all q-dimensional boundary chains, that is, the image of the homomorphism $\partial: C_{q+1} \to C_q$, is called the *group B_q of the q-dimensional boundaries* of $\mathfrak{C}$. For $q = n$, we have $B_n = 0$; for $q = 0$, we have $Z_0 = C_0$; the boundary of a zero-dimensional chain is always equal to zero.

12.2 Definition: In a chain complex $\mathfrak{C}$, the factor groups $Z_q/B_q = H_q = H_q(\mathfrak{C})$ are called the *q-dimensional homology groups of* $\mathfrak{C}$. If $H_q(\mathfrak{C}) = 0$ for every q, then the complex is said to be *exact*.

In particular, we have that $H_n = Z_n$ since $B_n = 0$, and $H_0 = C_0/B_0$ since $Z_0 = C_0$.

12.3 Definition: Let $\mathfrak{C}$ and $\mathfrak{D}$ be n-dimensional chain complexes. A chain mapping $f: \mathfrak{C} \to \mathfrak{D}$ is a *system of homomorphisms* $f_q: C_q \to D_q$ which are compatible with the boundary mapping, i.e., $\partial f_q = f_{q-1}\partial$ holds for all q.

A chain mapping f can be represented especially clearly by means of a diagram:

$$\begin{array}{ccccccccccccc} 0 & \xrightarrow{\partial} & C_n & \xrightarrow{\partial} & \cdots & \xrightarrow{\partial} & C_q & \xrightarrow{\partial} & C_{q-1} & \xrightarrow{\partial} & \cdots & \xrightarrow{\partial} & C_0 & \xrightarrow{\partial} & 0 \\ & & \downarrow f_n & & & & \downarrow f_q & & \downarrow f_{q-1} & & & & \downarrow f_0 & & \\ 0 & \xrightarrow{\partial} & D_n & \xrightarrow{\partial} & \cdots & \xrightarrow{\partial} & D_q & \xrightarrow{\partial} & D_{q-1} & \xrightarrow{\partial} & \cdots & \xrightarrow{\partial} & D_0 & \xrightarrow{\partial} & 0. \end{array}$$

The condition of boundary preservation asserts here that we can go from C_q to D_{q-1} either by the path through C_{q-1}, namely through $f_{q-1}\partial$ (first ∂, then f_{q-1}), or else by the path through D_q, namely through ∂f_q (first f_q, then ∂). Both paths yield the same homomorphism; in other words, the rectangle C_q, C_{q-1}, D_{q-1}, D_q is commutative. Note that for the sake of simplicity we have denoted both the boundary operations in $\mathfrak{C}$ and $\mathfrak{D}$ by ∂.

12.4 Theorem: *A chain mapping* $f\colon \mathfrak{C} \to \mathfrak{D}$ *induces in every dimension* q *a homomorphism* $f_*\colon H_q(\mathfrak{C}) \to H_q(\mathfrak{D})$.

The proof proceeds exactly as the corresponding considerations at the end of §8. We naturally obtain examples for chain complexes and chain mappings in the chain groups of a simplicial complex K with the boundary operator ∂; they yield the *chain complex* $\mathfrak{C}(\mathsf{K})$ *assigned to* K. But we will also have some other examples. Simplicial mappings $\varphi\colon \mathsf{K} \to \Lambda$ with the corresponding mappings of the chain groups furnish examples of chain mappings.—But also the cochain groups C^q of a simplicial complex with the coboundary operator form chain complexes as we see when we replace the dimension indices $0, \ldots, n$ in sequence by $n, \ldots, 0$. The cohomology groups $H^q(\mathsf{K})$ are then homology groups of this chain complex.

The concept formed in this section are even more general since the groups C_q are now arbitrary abelian groups, not necessarily free groups. This plays no role whatsoever in our development, but it is important for other purposes.

Two chain mappings $f, g\colon \mathfrak{C} \to \mathfrak{D}$ are said to be *homotopic* (more specifically, *chain homotopic* or *algebraically homotopic*) if for each q there exists a homomorphism $\boldsymbol{D}\colon C_q \to D_{q+1}$ with the operator property $\partial \boldsymbol{D} + \boldsymbol{D}\partial = g - f$. This relation of homotopy among all the chain mappings of $\mathfrak{C}$ into $\mathfrak{D}$ is reflexive, symmetric, and transitive. It leads to a partition of these mappings into mapping classes. Without any difficulty, we can carry over Theorem 11.2 together with its proof to homotopic chain mappings. However, we shall not need these facts in the sequel.

§ 13. Subcomplexes and Factor Complexes

13.1 Definition: Let $\mathfrak{C}$ be an n-dimensional chain complex with the chain groups C_q and the boundary operator ∂. An n-dimensional chain complex $\mathfrak{D}$, whose chain groups D_q for every q are subgroups of the C_q and whose boundary operator coincides with ∂ when applied to the elements of D_q, is called a *chain subcomplex*, or briefly a *subcomplex*, of $\mathfrak{C}$. We write $\mathfrak{D} \subset \mathfrak{C}$.

Without danger of misunderstanding, we denote the boundary operator in $\mathfrak{D}$ also by ∂.—Note that under the assumptions made, $\partial D_q \subset D_{q-1}$.—As above, let $\mathfrak{D} \subset \mathfrak{C}$ be n-dimensional chain complexes. For each q, we form the factor group C_q/D_q. Since $\partial D_q \subset D_{q-1}$, all q-dimensional chains of a residue class of C_q with respect to D_q have boundaries in the same residue class of C_{q-1} with respect to D_{q-1}. Therefore, we can define a unique boundary operator $\bar{\partial}$ among the factor groups C_q/D_q by applying the boundary operator ∂ to the representatives of a residue class. This boundary operator $\bar{\partial}$, determined by ∂, obviously satisfies the equation $\bar{\partial}^2 = 0$ and, hence, makes the sequence of factor groups into a chain complex. We now make the following definition.

13.2 Definition: Let $\mathfrak{D} \subset \mathfrak{C}$ be n-dimensional chain complexes with the boundary operator ∂. The factor groups C_q/D_q with the boundary operator $\bar{\partial}$, determined by ∂, also form an n-dimensional chain complex, called the *factor complex* $\mathfrak{C}/\mathfrak{D}$.

When no confusion is possible, we shall occasionally denote the boundary operator $\bar{\partial}$ simply by ∂.—Again, let $\mathfrak{D} \subset \mathfrak{C}$. We consider, for each q, the *inclusion mapping* $i: D_q \to C_q$, which assigns to each element of D_q the element itself. If we apply Theorem 12.4 of the preceding section to i, then we have the *inclusion homomorphism* induced by i,

$$i_*: H_q(\mathfrak{D}) \to H_q(\mathfrak{C}).$$

Whereas i is monomorphic, the same need not necessarily hold for i_*, as we will see later with some examples.—Furthermore, for each q, there exists a natural *projection mapping* j: $C_q \to C_q/D_q$, which assigns to each chain of C_q its residue class with respect to D_q. By virtue of Theorem 12.4 of the preceding section, j induces the *projection homomorphism* j_*,

$$j_*: H_q(\mathfrak{C}) \to H_q(\mathfrak{C}/\mathfrak{D}).$$

Whereas j is epimorphic, the same need not necessarily hold for j_*, as we shall also see from examples.

We apply these general considerations to simplicial complexes. Let K be a simplicial complex and let Λ be a subcomplex of K. Then, for the corresponding chain complexes $\mathfrak{C}(\mathsf{K})$ and $\mathfrak{C}(\Lambda)$, we have that $\mathfrak{C}(\Lambda) \subset \mathfrak{C}(\mathsf{K})$. We also consider the factor complex $\mathfrak{C}(\mathsf{K})/\mathfrak{C}(\Lambda)$, which will be briefly denoted as the *relative chain complex* $\mathfrak{C}(\mathsf{K}/\Lambda)$ of K with respect to Λ. Its chains are residue classes of chains from $\mathfrak{C}(\mathsf{K})$ with respect to the chain groups of $\mathfrak{C}(\Lambda)$. Its homology groups are called the *relative homology groups* $H_q(\mathsf{K}/\Lambda)$ *of* K *with respect to* Λ. They play an important role in the further development of homology theory. A relative q-dimensional cycle is represented by a q-dimensional chain of K, whose boundary lies in Λ; a relative q-dimensional boundary is represented by a chain of K, which is representable as the sum of a boundary from K and a chain from Λ. The inclusion mapping $i\colon \mathfrak{C}(\Lambda) \to \mathfrak{C}(\mathsf{K})$ and the projection mapping $j\colon \mathfrak{C}(\mathsf{K}) \to \mathfrak{C}(\mathsf{K}/\Lambda)$ induce the corresponding homomorphisms

$$i_*\colon H_q(\Lambda) \to H_q(\mathsf{K}), j_*\colon H_q(\mathsf{K}) \to H_q(\mathsf{K}/\Lambda).$$

In general, i_* is not monomorphic (and *a fortiori* not epimorphic); *there exist cycles which do not bound in* Λ *but do, however, bound in* K. As an example, let K be a circular disk and Λ its boundary; the generating cycle of $H_1(\Lambda)$ is homologous to zero in K.—In general, j_* is not epimorphic (and *a fortiori* not monomorphic); *there exist relative cycles of* K *with respect to* Λ *which are not the image under* j_* *of a cycle of* K. In the above example, the circular disk represents a 2-dimensional relative cycle of this sort.

For the relative homology groups $H_q(\mathsf{K}/\Lambda)$, there exist still other, formally somewhat simpler constructions which we shall utilize later. They depend only on the subcomplex $\mathsf{N} = \mathsf{K} - \Lambda$ and are introduced as follows: In every residue class of $C_q/D_q = C_q(\mathsf{K})/C_q(\Lambda)$, we may distinguish a unique representative, namely that chain for which all simplexes of Λ appear with the coefficient 0. We can conceive of them as chains on the subcomplex $\mathsf{N} = \mathsf{K} - \Lambda$. (Note that up to this point we have considered chains only on complexes.) If we wish to extend the boundary mappings also to these chains of N, then we must introduce for each simplex s_q of N a *modified boundary operator* ∂': $\partial' s_q$ applies only to those $(q-1)$-dimensional faces of s_q which occur in N, $\partial' s_q$ consists only of the "portion" of the usual boundary ∂s_q which lies on N. Thus, we obtain a representation of $\mathfrak{C}(\mathsf{K}/\Lambda)$ which depends only on N and which, therefore, will be denoted by $\mathfrak{C}(\mathsf{N})$. We have that $\mathfrak{C}(\mathsf{N})$ *is the chain complex assigned to a subcomplex* N. We can express this in the following theorem.

13.3 Theorem: *To each subcomplex* N *of a complex* K *there is assigned by virtue of the modified boundary formation* ∂' *a chain complex* $\mathfrak{C}(\mathsf{N})$ *for whose homology group the following holds:*

$$H_q(\mathsf{N}) \approx H_q(\text{Hu } \mathsf{N}/\text{Bd}\mathsf{N}).$$

As an example, we consider in an arbitrarily given simplicial complex K the subcomplex Φ^r which consists of a single r-dimensional simplex σ^r. Obviously, from among all the homology groups $H_q(\Phi^r)$, only $H_r(\Phi^r)$ can contain elements $\neq 0$, and $H_r(\Phi^r) \approx \mathsf{Z}$ is infinite cyclic. The subcomplex Φ^r is called an *r-dimensional elementary cell*, or, briefly, an *r-cell*. We shall deal in detail with cells in Part III (Cell Complexes). For an overall view, we place in juxtaposition the homology groups for the r-element, the r-sphere (§ 9), and the r-cell:

$$\begin{aligned} &r\text{-element } \mathsf{T}^r: (H_0, \ldots, H_r) = (\mathsf{Z}, 0, \ldots, 0, 0),\\ &r\text{-sphere } \textstyle\sum^r: (H_0, \ldots, H_r) = (\mathsf{Z}, 0, \ldots, 0, \mathsf{Z}),\\ &r\text{-cell} \quad \Phi^r: (H_0, \ldots, H_r) = (0, 0, \ldots, 0, \mathsf{Z}). \end{aligned}$$

We can now supplement Theorem 9.8 on closed orientable n-dimensional pseudomanifolds by a corresponding theorem on non-closed orientable n-dimensional pseudomanifolds $\mathsf{M} = \mathsf{M}^n$ since such M^n also possess homology groups in the sense of the last theorem. First of all, the previous discussion about $H_n(\mathsf{M}^n)$ and the fundamental cycle of M^n remains unchanged *verbatim*. In order to determine $H_0(\mathsf{M}^n)$, choose a vertex $\mathfrak{p}_0 \in \text{Bd } \mathsf{M}^n$. It is not contained in M^n. In Hu M^n, every vertex $\mathfrak{p} \in \mathsf{M}^n$ can be joined with $\mathfrak{p}_0$ by a polygonal arc; therefore, in Hu M^n, we have $\langle \mathfrak{p} \rangle - \langle \mathfrak{p}_0 \rangle \sim 0$. It follows from this that $\langle p \rangle \sim 0$ holds in M^n. Hence, $H_0(M^n) \approx 0$, and we can state the following theorem.

13.4 Theorem: *The statement of Theorem 9.8 is also valid for non-closed orientable n-dimensional pseudomanifolds* M^n *with the exception of the assertion about* $H_0(M^n)$; *moreover, in this case, we have that* $H_0(M^n) \approx 0$.

For later applications, we prepare the following two theorems which concern special subcomplexes, namely stars of a simplicial complex.

13.5 Theorem: *Let* $\Theta = \mathrm{St}\ p$ *be a vertex star, consisting of more than the vertex* p, *of a complex. Let* $H = \mathrm{Hu}\ \Theta$, $P = H - \Theta$. *Then the following hold:*

$$H_q(\Theta) \approx H_{q-1}(P) \quad \text{for} \quad q \geqq 2,$$

$$H_1(\Theta) \approx (k-1) \times Z, \quad H_0(\Theta) \approx 0,$$

where k is the number of components in P.

Here $(k-1) \times Z$ means the direct sum of $k-1$ summands of the type Z.—The *proof* depends on the fact that for $q \geqq 1$ a natural one-to-one relation holds between the q-simplexes $s_q = \langle pp_1 \ldots p_q \rangle$ of Θ and the $(q-1)$-simplexes $t_{q-1} = \langle p_1 \ldots p_q \rangle$ of P by means of which also the chain groups $C_q(\Theta)$ and $C_{q-1}(P)$ are related in one-to-one manner to one another. This relation preserves boundaries up to sign for $q \geqq 2$, since, upon the application of the modified boundary operator ∂', we have that $\partial' s_q = -\partial t_{q-1}$ is valid in Θ (as can easily be verified). Since the homology groups H_q depend only on the groups C_{q+1}, C_q and C_{q-1} and the bounding relations between them, the first assertion of the theorem is clear.—For $q = 1$, a special deduction is necessary. First $C_1(\Theta)$ and $C_0(P)$ are connected in a one-to-one fashion by the above relation; also, the boundary chains in $C_1(\Theta)$ correspond to the boundary chains in $C_0(P)$. For the cycles, however, the following holds: Whereas every chain of $C_0(P)$ is a cycle, a 1-dimensional chain from Θ of the form $\sum g_i \langle pq_i \rangle$, where q_i ranges over the points of P, is a cycle if and only if the coefficient sum $\sum g_i$ vanishes. If we choose in each component of P a point p_j ($j = 1, \ldots, k$), then the cycles $\langle p_j \rangle$ represent a basis for $H_0(\mathsf{P})$. Clearly, the chains $\langle pp_1 \rangle - \langle pp_j \rangle$ are cycles for $j = 2, \ldots, k$ and indeed a basis of $H_1(\Theta)$. The second part of the theorem is thus also proved.—That $H_0(\Theta) \approx 0$ follows from $\partial'(pp_i) = -\langle p \rangle$.

13.6 Theorem: *Suppose the complex* K *consists of an* $(r-1)$-*dimensional simplex* $\sigma = \sigma^{r-1}$ *and* $k(\geqq 1)$ r-*dimensional simplexes* $\varphi_i = \sigma_i^r$ $(i = 1, \ldots. k)$, *which are incident with* σ, *and of all faces of these simplexes. Then the following holds for the star* $\Theta = \mathrm{St}\ \sigma$: $H_r(\Theta) \approx (k-1) \times \mathsf{Z}$, $H_q(\Theta) \approx 0$ *for* $q \neq r$.

Proof: If we introduce suitably oriented simplexes s_{r-1} and $s_r^i (i = 1, \ldots, k)$ for σ and the σ_i, respectively, then the modified boundary homomorphism ∂' in Θ yields the bounding relations $\partial' s_r^i = s_{r-1}$. Clearly, the chains $s_r^1 - s_r^i$ for $i = 2, \ldots, k$ are cycles, and, indeed, a basis for the group $Z_r(\Theta) \approx H_r(\Theta)$, from which the assertion follows.

Our considerations from the beginning of this section about subcomplexes and factor complexes are now applied to cohomology groups. We limit ourselves for the sake of brevity to stating the results for the cohomology groups. Let Λ be a subcomplex of the simplicial complex K. Those cochains of $C^q = C^q(\mathsf{K})$ which have the value zero on Λ form a subgroup D^q. Since $\mathsf{K} - \Lambda$ is open in K, the coboundary of such a chain is again of the same sort, i.e., δ maps D^q into D^{q+1}. In this way, we obtain in the D^q a subcochain complex of the cochain complex of the C^q. The corresponding homology groups are called the *relative cohomology groups* $H^q(\mathsf{K}/\Lambda)$ *of* K *with respect to* Λ. The corresponding inclusion mapping is here denoted by j^*, corresponding to the mapping j_*,

$$j^*: H^q(\mathsf{K}/\Lambda) \to H^q(\mathsf{K}).$$

The factor complex from the groups C^q/D^q, obviously, yields the cohomology groups of Λ. The corresponding projection mapping is

$$i^*: H^q(\mathsf{K}) \to H^q(\Lambda).$$

In general, j^* is not monomorphic (and *a fortiori* not epimorphic); *there exist non-zero homologous cocycles in* $\mathfrak{C}(\mathsf{K}/\Lambda)$ *which are coboundaries in* K. In general, i^* is not epimorphic (and *a fortiori* not monomorphic); *there exist cocycles in* Λ *which have no pre-image in* K.

§ 14. The Boundary Operator

Let $\mathfrak{D} \subset \mathfrak{C}$ be two chain complexes. To the two homomorphisms i_* and j_* which connect the homology groups $H_q(\mathfrak{D})$, $H_q(\mathfrak{C})$ and $H_q(\mathfrak{C}/\mathfrak{D})$ with one another, we now introduce another homomorphism,

$$\partial_*: H_q(\mathfrak{C}/\mathfrak{D}) \to H_{q-1}(\mathfrak{D}).$$

In contrast to i_* and j_*, the homomorphism ∂_* lowers dimension and is associated with the usual boundary formation ∂, but must not be confused with ∂. And ∂_* is defined as follows. Let $h \in H_q(\mathfrak{C}/\mathfrak{D})$ be a relative homology class. A relative cycle from h is a residue class with respect to D_q of a chain $c = c_q$ from $\mathfrak{C}_q$ whose boundary $\partial c_q = d_{q-1}$ lies in D_{q-1}. Since $\partial d_{q-1} = \partial^2 c_q = 0$, we have that d_{q-1} is a cycle from D_{q-1}. Let h_1 be that homology class of $H_{q-1}(\mathfrak{D})$ to which $d = d_{q-1}$ belongs. We then set $\partial_* h = h_1$.—It must be shown that h_1 remains fixed when we replace c by an arbitrary representative of another relative cycle from h, say c'. Since c and c' differ by a boundary relative to D_q, we have $c' = c + \partial c^0 + d^0$ with $c^0 \in C_{q+1}$ and $d^0 \in D_q$. It follows that $d' = \partial c' = \partial c + \partial d^0$, and $d' = d + \partial d^0$; hence, $d' \sim d$ in $\mathfrak{D}$, which is the desired result.—That ∂_* yields a homomorphism of $H_q(\mathfrak{C}/\mathfrak{D})$ into $H_{q-1}(\mathfrak{D})$ is clear from the definition. With the aid of ∂_* we form a sequence of homomorphisms,

$$\ldots \to H_q(\mathfrak{D}) \xrightarrow{i_*} H_q(\mathfrak{C}) \xrightarrow{j_*} H_q(\mathfrak{C}/\mathfrak{D}) \xrightarrow{\partial_*} H_{q-1}(\mathfrak{D}) \to \ldots,$$

which is called the *homology sequence of the pair* $\mathfrak{C}$, $\mathfrak{D}$. Extending in both directions, it begins with

$$0 \to H_n(\mathfrak{D}) \to H_n(\mathfrak{C}) \to H_n(\mathfrak{C}/\mathfrak{D}) \to H_{n-1}(\mathfrak{D}) \to \ldots$$

and ends with

$$\ldots \to H_0(\mathfrak{D}) \to H_0(\mathfrak{C}) \to H_0(\mathfrak{C}/\mathfrak{D}) \to 0.$$

The following important theorem holds.

14.1 Theorem: *The homology sequence of a pair* $\mathfrak{D} \subset \mathfrak{C}$ *of chain complexes is exact.*

This statement means, as a slight generalization of Definition 12.2, that the image of every homomorphism of the sequence is equal to the kernel of the subsequent homomorphism.—The *proof* decomposes accordingly into six parts: First is to show that the sequence is exact at $\mathfrak{C}$, i.e., that (1) $j_* \circ i_* = 0$, in other words, that the image of i_* is contained in the kernel of j_*, $\operatorname{Im} i_* \subset \operatorname{Ker} j_*$, and that (2) $\operatorname{Im} i_* \supset \operatorname{Ker} j_*$. From (1) and (2), it follows that $\operatorname{Im} i_* = \operatorname{Ker} j_*$. Then it is to be shown correspondingly (3), (4) that the sequence is exact at $\mathfrak{C}/\mathfrak{D}$ and then (5), (6) that it is exact at $\mathfrak{D}$. It suffices to prove this assertion about homology classes for suitable representatives. In the following proof, ∂ always denotes the boundary operator in $\mathfrak{C}$, which for chains from $\mathfrak{D}$ coincides with the boundary operator in $\mathfrak{D}$. The six parts of the proof now follow.

(1) It follows immediately from $j \circ i = 0$ that $j_* \circ i_* = 0$.—(2) Let z be a cycle from $\mathfrak{C}$ which as a relative cycle is homologous to zero, that is, $z = \partial c + d$, with $c \in \mathfrak{C}$, $d \in \mathfrak{D}$. Then we assert that there exists a cycle $y \in \mathfrak{D}$ for which $z \sim y$ in $\mathfrak{C}$. In fact, we need only take $y = d$. From $z = \partial c + d$, it follows by taking the boundary that d is a cycle, and hence that $z \sim y$.—(3) For a cycle $z \in \mathfrak{C}$, we have $\partial z = 0$. If we form the ∂_*-image of the homology class of z according to the procedure above, then $d = \partial z = 0$.—(4) As in the above definition, let $d = \partial c \in \mathfrak{D}$ for some $c \in \mathfrak{C}$ be a relative cycle and let d be homologous to zero in $\mathfrak{D}$, i.e., $d = \partial d^0$. We assert that there exists a cycle $z \in \mathfrak{C}$, which, as a relative cycle, is homologous to c. In fact, we can choose $z = c - d^0$. Then, we have $\partial c = \partial d^0$ or $\partial(c - d^0) = 0$; therefore, z is a cycle, and $c - d^0$ is obviously relatively homologous to c.—(5) Let $c \in \mathfrak{C}$ represent a relative cycle and let $\partial c = d \in \mathfrak{D}$ be its ∂_*-image. Obviously, d is homologous to zero in $\mathfrak{C}$.—(6) Let $z \in \mathfrak{D}$ be homologous to zero in $\mathfrak{C}$, $z = \partial c$ with $c \in \mathfrak{C}$. It is then asserted that z lies in a homology class which is a ∂_*-image. In fact, c represents a relative cycle, and the procedure of the above definition yields as its ∂_*-image the class of $\partial c = z$.

The applications of these concepts and theorems to simplicial complexes are immediate: Let $\mathsf{\Lambda}$ be a subcomplex of the simplicial complex K. There exists a boundary operator δ_* and a coboundary operator δ^*, which represent the homomorphisms

$$\partial_*: H_q(\mathsf{K}/\mathsf{\Lambda}) \to H_{q-1}(\mathsf{\Lambda}),$$

$$\delta^*: H^{q-1}(\mathsf{\Lambda}) \to H^q(\mathsf{K}/\mathsf{\Lambda})$$

and lead to the *homology sequence of* $\mathsf{K}/\mathsf{\Lambda}$,

$$\ldots H_q(\mathsf{\Lambda}) \xrightarrow{i_*} H_q(\mathsf{K}) \xrightarrow{j_*} H_q(\mathsf{K}/\mathsf{\Lambda}) \xrightarrow{\partial_*} H_{q-1}(\mathsf{\Lambda}) \to \ldots,$$

and to the *cohomology sequence of* $\mathsf{K}/\mathsf{\Lambda}$,

$$\ldots \to H^{q-1}(\mathsf{\Lambda}) \xrightarrow{\delta^*} H^q(\mathsf{K}/\mathsf{\Lambda}) \xrightarrow{j^*} H^q(\mathsf{K}) \xrightarrow{i^*} H^q(\mathsf{\Lambda}) \to .$$

Both of these sequences are exact.

Chapter 4

FREE CHAIN COMPLEXES

§ 15. Modules and Dual Modules

We now specialize our chain complexes somewhat in order to make them more suitable for the anticipated applications to simplicial complexes. Namely, we assume that the groups C_q are free abelian groups or, as we shall say, *modules*, and indeed modules of finite rank. Such a module of *rank* α possesses a basis $s_q{}^i$ ($i = 1, \ldots, \alpha$) with the property that every chain $c_q \in C_q$ is representable uniquely in the form $c_q = \sum g_i s_q{}^i$ with integer g_i, summed over i. We combine the elements of such a basis into a column vector $\mathfrak{c}_q$.—The index q is brought in only for purposes of denoting.

Naturally, there are many bases for C_q. If $\mathfrak{t}_q$ is another basis, also represented as an α-membered column vector, then $\mathfrak{c}_q$ can be expressed linearly and homogeneously in terms of $\mathfrak{t}_q$ in matrix form, that is, by $\mathfrak{c}_q = P\mathfrak{t}_q$ with integer α-rowed quadratic matrix P. There is also a representation in the reverse direction, $\mathfrak{t}_q = Q\mathfrak{c}_q$, with a matrix Q of the same form. It follows that $\mathfrak{c}_q = PQ\mathfrak{c}_q$; therefore, because of the uniqueness of the representation, $PQ = E$, the unit matrix. It follows further for the determinants that $|PQ| = 1$, $|P| = \pm 1$, $Q = P^{-1}$; hence, P and Q are *unimodular matrices*, i.e., integer matrices with determinant ± 1 with integer inverses. Two bases $\mathfrak{c}_q$ and $\mathfrak{t}_q$ of C_q are therefore interrelated by unimodular transformation matrices.

Upon introduction of the *dual module* C^q to C_q we can express ourselves briefly since the considerations run parallel to the investigation of cochains in §10. We consider homomorphisms $c^q\colon C_q \to \mathbf{Z}$, where Z denotes as always the additive group of integers. And c^q is also called a cochain and assigns to each chain $c_q \in C_q$ an integer which is denoted by (c_q, c^q) and is called the *Kronecker product* of c_q and c^q. The c^q themselves form a group C^q and both of the distributive laws are valid:

$$(c_q{}^1 + c_q{}^2, c^q) = (c_q{}^1, c^q) + (c_q{}^2, c^q),$$
$$(c_q, c_1{}^q + c_2{}^q) = (c_q, c_1{}^q) + (c_q, c_2{}^q).$$

The first holds because c^q is a homomorphism and the second is the definition of the addition of two homomorphisms. To the chains $s_q{}^i$ $(i = 1, \ldots, \alpha)$ of a basis of C_q we can let correspond cochains $s_j{}^q$ $(j = 1, \ldots, \alpha)$ by the convention that $(s_q{}^i, s_j{}^q) = \delta_{ij} = 1$ or 0 according as $i = j$ or $i \neq j$. If we also combine the $s_i{}^q$ into an α-membered column vector $\mathfrak{c}^q$, then the following, written in matrix form, is valid:

$$(\mathfrak{c}_q, \mathfrak{c}^{q\mathsf{T}}) = E,$$

where T denotes the transpose of a matrix, i.e., the reflection in the principal diagonal. The $s_j{}^q$ obviously form a basis for C^q. Therefore C_q and C^q have the same order α; they are related by the Kronecker product (c_q, c^q) in a bilinear manner.—If $\mathfrak{t}_q$ is another basis of C_q, $\mathfrak{c}_q = P\mathfrak{t}_q$ with unimodular matrix P, then $\mathfrak{t}_q = P^{-1}\mathfrak{c}_q$ and if $\mathfrak{t}^q$ is another basis for C^q, then $\mathfrak{c}^q = Q\mathfrak{t}^q$, and $\mathfrak{t}^{q\mathsf{T}}Q^{-\mathsf{T}}$, where $Q^{-\mathsf{T}} = (Q^{-1})^{\mathsf{T}} = (Q^{\mathsf{T}})^{-1}$ is the matrix contragredient to Q, then, according to the rules of the matrix calculus, the following is valid:

$$(\mathfrak{t}_q, \mathfrak{t}^{q\mathsf{T}}) = (P^{-1}\mathfrak{c}_q, \mathfrak{c}^{q\mathsf{T}}Q^{-\mathsf{T}}) = P^{-1}(\mathfrak{c}_q, \mathfrak{c}^{q\mathsf{T}})Q^{-\mathsf{T}} = P^{-1}Q^{-\mathsf{T}}.$$

This is equal to the unit matrix if and only if $Q = P^{-\mathsf{T}}$; in this case, $\mathfrak{t}^q$ is the dual basis to $\mathfrak{t}_q$.—We note emphatically the following theorem.

15.1 Theorem: *To every basis $\mathfrak{c}_q$ of C_q there exists a uniquely determined dual basis $\mathfrak{c}^q$ of C^q which is determined by $(\mathfrak{c}_q, \mathfrak{c}^{q\mathsf{T}})$.*

§ 16. Mappings and Dual Mappings

Let $f: C_q \to D_q$ be a *mapping* and indeed a homomorphism of the module C_q of order α into the module D_q of order β. If we refer to bases $\mathfrak{c}_q$ of C_q and $\mathfrak{u}_q$ of D_q, then f can be expressed by means of an equation system $f(\mathfrak{c}_q) = F_q\mathfrak{u}_q$, in which the mapping matrix is an integer matrix with α rows and β columns, which in turn obviously determines f completely. If other bases of C_q and D_q are given, say $\mathfrak{c}_q = P\mathfrak{t}_q$ and $\mathfrak{u}_q = Q\mathfrak{v}_q$, then $f(P\mathfrak{t}_q) = F_qQ\mathfrak{v}_q$, $f(\mathfrak{t}_q) = (P^{-1}F_qQ)\mathfrak{v}_q$. The new matrix characterizing f is therefore $P^{-1}F_qQ$. We go briefly into the special case when $D_q = C_q$ holds, i.e., when f is a mapping of C_q into itself. In this case, F_q is quadratic and the matrices characterizing f are F_q and $P^{-1}F_qP$. By the *trace* $\mathrm{Tr}(F)$ of an α-membered quadratic matrix F we understand the sum of the diagonal elements of F; $\mathrm{Tr}(F)$ is accordingly the factor which when multiplied by $(-1)^{\alpha-1}$ is the coefficient of $x^{\alpha-1}$ in the characteristic polynomial $|F - xE|$ of F. We assert that F_q and $P^{-1}F_qP$ have the same trace. Namely, both of the corresponding characteristic polynomials coincide on account of

$$|P^{-1}FP - xE| = |P^{-1}(F - xE)P| = |P|^{-1}|F - xE||P|$$

and hence also all of their coefficients coincide. We can thus note for later purposes that *to a mapping of a module C_q into itself, $f: C_q \to C_q$, there corresponds a well-defined trace* $\mathrm{Tr}(f)$ *which depends only on f and not on a special basis of C_q.*

For f there can be defined a *dual mapping* $f^\bullet: D^q \to C^q$ of the dual modules. In the sequel, for reasons of symmetry, we denote the mapping f by $f_\bullet$. In order to define the image $f^\bullet d^q$ of a cochain $d^q \in D_q$, we prescribe the value with which an arbitrary chain $c_q \in C_q$ endows $f^\bullet d^q$:

$$(c_q, f^\bullet d^q) = (f_\bullet c_q, d^q) \quad \text{(Stokes's Formula).}$$

Here, $f^\bullet d^q$ is a homomorphism of C_q into Z, which is apparent if we replace c_q by $c_q^1 + c_q^2$ and observe the effect on the right member.

To the sum $d_1{}^q + d_2{}^q$ of two elements of D^q there corresponds upon application of $f^\bullet$, on account of

$$\begin{aligned}(c_q, f^\bullet(d_1{}^q + d_2{}^q)) &= (f_\bullet c_q, d_1{}^q + d_2{}^q)\\ &= (f_\bullet c_q, d_1{}^q) + (f_\bullet c_q, d_2{}^q)\\ &= (c_q, f^\bullet d_1{}^q) + (c_q, f^\bullet d_2{}^q),\end{aligned}$$

the sum $f^\bullet d_1{}^q + f^\bullet d_2{}^q$. Thus, $f^\bullet$ is actually a homomorphism, $f^\bullet: D^q \to C^q$. Also $f^\bullet$ can be described by means of a mapping matrix, $f^\bullet \mathfrak{u}^q = F^q \mathfrak{c}^q$; F^q has β rows and α columns. If we substitute in $(\mathfrak{c}_q, f^\bullet(\mathfrak{u}^{q\mathsf{T}})) = (f_\bullet(\mathfrak{c}_q), \mathfrak{u}^{q\mathsf{T}})$ the mapping matrices, then it follows that $(\mathfrak{c}_q, \mathfrak{c}^{q\mathsf{T}} F^{q\mathsf{T}}) = (F_q \mathfrak{u}_q, \mathfrak{u}^{q\mathsf{T}})$, $(\mathfrak{c}_q, \mathfrak{c}^{q\mathsf{T}})F^{q\mathsf{T}} = F_q(\mathfrak{u}_q, \mathfrak{u}^{q\mathsf{T}})$, $F^q = F_q{}^{\mathsf{T}}$. We note the result in the following theorem.

16.1 Theorem: *To every mapping $f_\bullet$: $C_q \to D_q$ of two modules there exists a uniquely determined dual mapping $f^\bullet$: $D^q \to C^q$ of the dual modules. Upon fixing the dual bases in C_q and C^q as well as in D_q and D^q the corresponding mapping matrices arise from one another by transposition.*

Examples of modules and dual modules are formed by the chain groups C_q and the cochain groups C^q of a simplicial complex. We have met mappings and dual mappings in the boundary construction ∂ and in the coboundary construction δ, further in the chain mappings and cochain mappings $f_\bullet$ and $f^\bullet$ induced by the simplicial mappings f. In particular, the formula 10.3 for the coboundary δ and the formula 10.8 for the mapping $\varphi^\bullet$ are special cases of the above Stokes's formula.—We must now investigate somewhat more exactly the structure of a mapping f: $C_q \to D_q$.

16.2 Theorem: *Let C and D be modules of rank α and β, respectively, and let f: $C \to D$ be a homomorphism. There exist bases c^i of C $(i = 1, \ldots, \alpha)$ and d^j of D $(j = 1, \ldots, \beta)$ such that $f(c^i) = e_i d^i$ for $i = 1, \ldots, r$ and $f(c^i) = 0$ for $i = r + 1, \ldots, \alpha$ holds for positive integers e_i having the property that $e_1 \mid e_2 \mid \ldots \mid e_r$. Such bases can be constructed in a finite number of steps; they are called canonical bases* *for f. And r is called the rank and the e_i are called the elementary divisors of f.*

Here, $e_1 \mid e_2$ means that the integer e_1 is a divisor of the integer e_2; the numbers $e_1, \ldots, e_r$ in the theorem form a division chain in which each number is a divisor of the immediately following one. As a *proof*, we give a construction for obtaining the canonical bases, which consists of several parts. If the c^i are any basis of C and the d^j any basis of D, then f is represented by the mapping equations $f(c^i) = \sum x_{ij} d^j$, summed over j. The x_{ij} are called the *coordinates* of $f(c^i)$ with respect to the basis of the d^j. If all $x_{ij} = 0$, then the present bases are already canonical. Otherwise there exists at least one $x_{ij} \neq 0$. Then there also exist bases with an $x_{ij} > 0$, as we see if we replace c^i by $-c^i$ if necessary. We now consider all bases c^i of C and all bases d^j of D and the totality of $x_{ij} > 0$ which appear in this connection. Let the minimum of all these positive integers be $x_{ij} = e_1 > 0$. By interchanging c^1 with c^i and d^1 with d^j we can attain an operation which represents a change of basis for which precisely $x_{11} = e_1$. Thus in all mapping equations occurring for arbitrary bases of C and D every positive coordinate is $\geqq x_{11} = e_1$.

We now assert (1) that all x_{1j} are divisible by $x_{11} = e_1$, so that they have the form $x_{1j} = e_1 q_{1j}$. To prove this, we divide x_{1j} by e_1 so that $x_{1j} = e_1 q_{1j} + r$ with integer q_{1j} and r with $0 \leqq r < e_1$. The equation $f(c^1) = e_1 d^1 + \ldots + x_{1j} d^j + \ldots$ goes over upon the replacement of d^1 by $d_*^1 = d^1 \to q_{1j} d^j$ which represents a basis transformation into $f(c^1) = e_1 d_*^1 + \ldots + r d^j + \ldots$. On account of the minimal choice of $x_{11} = e_1$ it follows that $r = 0$ so that $x_{1j} = e_1 q_{1j}$ is in fact divisible by e_1.—We assert further (2) that also all x_{i1} are divisible by $x_{11} = e_1$, that they therefore have the form $x_{i1} = e_i q_{i1}$. To prove this, we divide x_{i1} by e_1 so that $x_{i1} = e_1 q_{i1} + r$ with integers q_{i1} and r with $0 \leqq r < e_1$. The equation $f(c^i) = x_{i1} d^1 + \ldots$ goes over upon the replacement of c^i by $c_*^i = c^i - q_{i1} c^1$ which again represents a basis transformation into $f(c^i) = r d^1 + \ldots$. On account of the minimal choice of $x_{11} = e_1$ it follows that $r = 0$ so that $x_{i1} = e_1 q_{i1}$ is in fact divisible be e_1.

If now in (3) we replace d^1 by $d^1 + \sum q_{1j} d^j$, which again represents a basis transformation, summed over $j = 2, \ldots, \beta$, then all coordinates of $f(c^1)$ vanish except the first one which is equal to e_1. Finally, if we replace, for $i = 2, \ldots, \alpha$, the basis element c_i by $c^i - q_{i1}c^1$, which also represents a basis transformation, then there result new mapping equations which have the form

$$\begin{cases} f(c^1) = e_1 d^1, \\ f(c^i) = \sum_{j=2}^{\beta} x_{ij} d^j \quad \text{for } i = 2, \ldots, \alpha. \end{cases}$$

It should be clear that these c^i, d^j and x_{ij} now have other values than at the beginning of the proof.

We assert that now all x_{ij} ($i = 2, \ldots, \alpha$; $j = 2, \ldots, \beta$) are divisible by e_1. Namely, if we replace c^1 by $c^1 + c^i$, which is again a basis transformation, then $f(c^1 + c^i) = e_1 d^1 + \ldots + x_{ij} d^j + \ldots$ and here, according to (1), x_{ij} is divisible by e_1. This concludes the first part of our construction.

The subsequent considerations refer only to the submodule C_1 of C which is generated by $c^2, \ldots, c^\alpha$ and the submodule D_1 of D which is generated by $d^2, \ldots, d^\beta$. Now f maps C_1 into D_1 and we apply the same procedure to this mapping that we applied above to C and D. For the bases of C_1 and D_1 just mentioned all the coordinates x_{ij} of the mapping equations are divisible by e_1 as we established at the end of the preceding paragraph. Therefore the same also holds upon fixing arbitrary bases of C_1 and D_1 because, as we saw at the beginning of this section, the mapping equations experience thereby a linear transformation. Thus the minimal number $e_2 > 0$, which appears in the application of our procedure to C_1 and D_1, is also divisible by e_1. The procedure yields mapping equations of the form

$$f(c^1) = e_1 d^1, f(c^2) = e_2 d^2,$$
$$f(c^i) = \sum x_{ij} d^j, \; i = 3, \ldots, \alpha, \text{ summed over } j = 3, \ldots, \beta.$$

In this connection, $e_1 \mid e_2$, and $e_2 \mid x_{ij}$ for $i, j \geqq 3$. The second part of our construction is thus completed.

The extension of the procedure terminates after a certain number $r \leqq \alpha, \beta$ of steps in the sense that all the remaining $x_{ij} = 0$. Thus the theorem is proved up to the assertion of the constructibility of the canonical bases in a finite number of steps.

Our construction requires in fact in each of its steps the choice of a minimum from the infinite set of positive numbers and is to this extent non-constructive, i.e., in general not accomplishable in a finite number of steps. This method was based on a minimum argument which was used only because of complexity of the proof; however, the proof can easily be modified so that the procedure becomes constructive. To this end, we need only, for instance in the first part, starting with arbitrary bases of C and D and $x_{11} > 0$, to apply the basis transformations (1), (2) and (3) one after the other and thereby to decrease the magnitude of x_{11} step by step. If this is no longer possible by means of (1), (2) and (3), then the canonical bases have been attained. This completes the proof of Theorem 16.2.—The $\alpha \times \beta$ mapping matrix of the canonical bases has the appearance sketched in Fig. 6: r non-vanishing natural

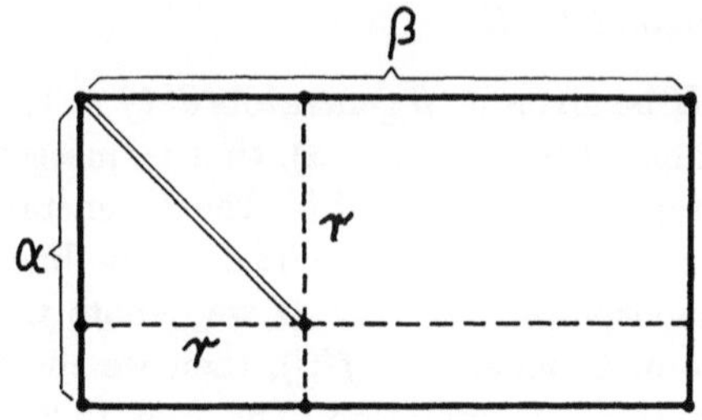

Fig. 6

numbers $e_1, \ldots, e_r$ only stand in the principal diagonal, designated by the double lines, where the $e_1, \ldots, e_r$ form a division chain; all the other matrix elements are zero.— The canonical bases of f are not at all uniquely determined by f, but we shall immediately see that the rank and the elementary divisors of f depend only on f and not on the special canonical bases.

If we now write the mapping f in the form $f_{\bullet}: C_q \to D_q$ and we adjoin the dual mapping $f^{\bullet}: D^q \to C^q$, then Theorem 16.1 asserts that upon application of dual bases the corresponding mapping matrices arise from one another by transposition. In this connection, the principal diagonal remains invariant (also see Fig. 6), and we can establish the following theorem.

16.3 Theorem: *The mapping $f_{\bullet}: C_q \to D_q$ and the dual mapping $f^{\bullet}: D^q \to C^q$ have the same rank and the same elementary divisors.*

With Theorem 16.2 we have at the same time proved the most essential part of the *fundamental theorem on finitely generated abelian groups* which we shall state as follows.

16.4 Theorem: *An abelian group H with a finite number of generators is the direct sum of a finite number $\wp$ of infinite cyclic groups and a finite number of finite cyclic groups of orders $e_1 \mid e_2 \mid \ldots \mid e_r$. The number $\wp$ is called the Betti number (or the rank) and the $e_1, \ldots, e_r$ are called the torsion numbers of H. Moreover, $\wp$ and $e_1, \ldots, e_r$ are uniquely determined by H.*

Proof: Let H be given by β generators d^j $(j = 1, \ldots, \beta)$ and by α so-called relations c^i $(i = 1, \ldots, \alpha)$, that is, integer linear forms in the d^j, which represent the zero of H. The d^j generate the module D, the c^i generate the module C; by virtue of the linear forms c^i, C is mapped homomorphically into D. If we denote this mapping by f and the image of C under f by $f(C)$, then we therefore have $H \approx D/f(C)$.—With this interpretation, Theorem 16.2 yields the representation of H as the direct sum of cyclic groups: $d^1, \ldots, d^r$ are the generators of the finite cyclic direct summands. In this connection, we remark that in Theorem 16.2 the first among the e_i can be equal to 1. But here we consider only the $e_i > 1$ which occur from a certain index on which we shall assume to be the case without an explicit renaming. The Betti number $\wp$ of H is equal to $\beta - r$.—Also if H is to be given by an infinite number of relations, for instance by all the representations of the zero, our method remains applicable and yields additionally the information that for β generators at most β suitable relations always suffice to define H.

It only remains to show that H uniquely determines the Betti number $\wp$ and the torsion numbers, $e_1, \ldots, e_r$. Obviously, $\wp$ is the maximal number of linearly independent elements in H; it is therefore uniquely determined by H. We point out how one recognizes the same thing for the torsion numbers. The elements of finite order of H form a finite subgroup T of H, the *torsion group of* H. For every prime number p, let $N(\nu)$ be the number, uniquely determined by H, of solutions $t \in T$ of the equation $p^\nu t = 0$. Further, let $n(\nu)$ be the number of the torsion numbers e_i which are divisible by p^ν. We recognize easily that $N(\nu)$ is equal to p to the power $n(1) + \ldots + n(\nu)$. Therefore, $n(1)$ is determined by $N(1)$, $n(2)$ is determined by $n(1)$ and $N(2)$, and so forth. Since the e_i are determined by the values of $n(\nu)$ for all prime numbers, they are thus also uniquely determined by H itself. This completes the proof of the fundamental theorem.

§ 17. Free Chain Complexes. Canonical Bases

We now consider free chain complexes $\mathfrak{C}$:

$$0 \xrightarrow{\partial} C_n \xrightarrow{\partial} \cdots \xrightarrow{\partial} C_q \xrightarrow{\partial} \cdots \xrightarrow{\partial} C_0 \xrightarrow{\partial} 0,$$

that is, for which all the C_q are free groups, i.e., modules. The mapping matrices of the boundary operator $\partial: C_q \to C_{q-1}$ referred to the bases $\mathfrak{c}_q$ of C_q are here called *bounding matrices* R_q; we have $\partial \mathfrak{c}_q = R_q \mathfrak{c}_{q-1}$; R_q has α_q rows and α_{q-1} columns; we set $\alpha_q = 0$ for $q > n$ and $q < 0$. The operator equation $\partial^2 = 0$ asserts here that $\partial^2 \mathfrak{c}_q = \partial R_q \mathfrak{c}_{q-1} = R_q \partial \mathfrak{c}_{q-1} = R_q R_{q-1} \mathfrak{c}_{q-2} = 0$, that therefore $R_q R_{q-1} = 0$. The mapping dual to ∂ is $\delta: C^q \leftarrow C^{q-1}$. It generates the chain complex dual to $\mathfrak{C}$,

$$0 \xleftarrow{\delta} C^n \xleftarrow{\delta} \cdots \xleftarrow{\delta} C^q \xleftarrow{\delta} \cdots \xleftarrow{\delta} C^0 \xleftarrow{\delta} 0.$$

If we choose for these modules bases $\mathfrak{c}^q$ which are dual to $\mathfrak{c}_q$, then we have $\delta \mathfrak{c}^{q-1} = R_q^{\mathsf{T}} \mathfrak{c}^q$. In the sequel, the rank r_q of $\partial: C_q \to C_{q-1}$, which is equal to the rank of R_q, and also the elementary divisors of this mapping, which are equal to those of R_q, play a particular role. The rank and elementary divisors of R_q^{T} are equal to those of R_q.

We now give an especially simple simultaneous system of bases for all C_q of $\mathfrak{C}$, the so-called *canonical bases for* $\mathfrak{C}$, which are a generalization of the canonical bases for a single mapping $f\colon C \to D$ as we considered them in Theorem 16.2. The corresponding bounding matrices are called *canonical bounding matrices*. From them we will derive along the way the important structure properties of the homology and cohomology groups.

17.1 Theorem: *Canonical bases can be introduced in the chain groups C_q of a free chain complex $\mathfrak{C}$ as follows: The basis of C_q $(q = 0, \ldots, n)$ consist of $\alpha_q = r_q + \wp_q + r_{q+1}$ elements of the form*

$$a_q^{i_q}, b_q^{i_q}, c_q^{i_q+1}$$

for $i_q = 1, \ldots, r_q$ and $j_q = 1, \ldots, \wp_q$, where $\wp_q$ denotes the q-dimensional Betti number of $\mathfrak{C}$ and r_q the rank of the bounding mapping, $\partial\colon C_q \to C_{q-1}$. We must set $r_{n+1} = r_0 = 0$, for $q = n$ there are no basis elements of the form $c_n^{i_n+1}$, and for $q = 0$ there are no basis elements of the form $a_0^{i_0}$. The boundary operations for these bases read for $q = 1, \ldots, n$ as follows:

$$\partial a_q^{i_q} = e_{i_q}^{(q)} c_{q-1}^{i_q}, \; \partial b_q^{j_q} = 0, \; \partial c_q^{i_q+1} = 0,$$

where the $e_{i_q}^{(q)}$ are the elementary divisors of the mapping $\partial\colon C_q \to C_{q-1}$ and form the division chain $e_1^{(q)} \mid e_2^{(q)} \mid \ldots \mid e_{r_q}^{(q)}$.

It is suggested that the reader compare the statement of this theorem with the adjoined sketch (see Fig. 7) for the case of dimension $n = 3$; this theorem is indeed somewhat involved but not especially difficult either as far as the actual assertion is concerned or as far as the attached proof is concerned. There the bounding matrices are indicated. In the left and in the upper entry stand the

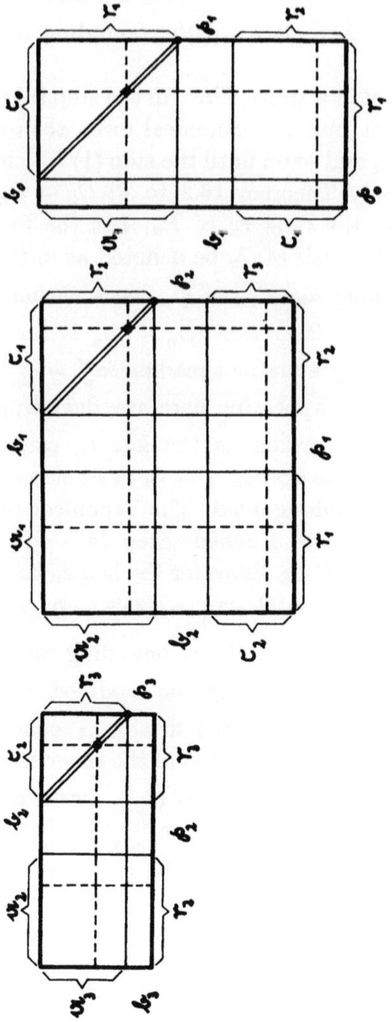

Fig. 7

basis elements, to the right and below stand the "side lengths," i.e., the numbers of the basis elements. In the adjacent diagonals designated by double lines stand the elementary divisors $e_{i_q}^{(q)}$ which are all $\neq 0$, and in all other places of the bounding matrix there stand zeros.

We carry out the *proof* in n steps. First, in the step (n) the bounding matrix R_n is brought into the canonical form, then in the step ($n-1$) the matrix R_{n-1} and so on until the step (1) which canonizes R_1.—Step (n): We apply Theorem 16.2 to $\partial \colon C_n \to C_{n-1}$. The r appearing there is here the rank r_n of R_n. Let the first r_n basis elements of the canonical basis of C_n be denoted as in the assertion by $a_n^{i_n}$ and the remaining $\alpha_n - r_n = \bar{\wp}_n$ basis elements by $b_n^{j_n}$. It will turn out somewhat later that $\bar{\wp}_n = \wp_n$ is the n-th Betti number of $\mathfrak{C}$. The boundary equations read as $\partial a_n^{i_n} = e_{i_n}^{(n)} c_{n-1}^{i_n}$. The basis elements $c_{n-1}^{i_n}$ of C_{n-1} appearing here are, deviating from the notation in Theorem 16.2, taken as the *last* r_n canonical basis elements of b_{n-1}, whereas the first $\alpha_{n-1} - r_n$ basis elements of C_{n-1} remain for the moment undesignated. The canonical form of the n-th bounding matrix is thus reached. Step ($n-1$): We apply Theorem 16.2 to $\partial \colon C_{n-1} \to C_{n-2}$. Since for the last r_n basis elements $c_{n-1}^{i_n}$ of C_{n-1} we have that $\partial e_{i_n}^{(n)} c_{n-1}^{i_n} = \partial^2 a_n^{i_n} = 0$ and therefore $\partial c_{n-1}^{i_n} = 0$ holds, the last r_n rows of the bounding matrix already vanish, so that the $c_{n-1}^{i_n}$ need no more be modified or taken into account in the canonization. Let the first r_{n-1} canonical basis elements of C_{n-1} arising upon application of the theorem be denoted by $a_{n-1}^{i_{n-1}}$, the next $\alpha_{n-1} - r_n - r_{n-1} = \bar{\wp}_{n-1}$ basis elements by $b_{n-1}^{j_{n-1}}$. The boundary equations now read $\partial a_{n-1}^{i_{n-1}} = e_{i_{n-1}}^{(n-1)} c_{n-2}^{i_{n-2}}$. Let the basis elements $c_{n-2}^{i_{n-1}}$ appearing here be taken, deviating from Theorem 16.2, as the *last* r_{n-1} canonical basis elements of C_{n-2}, whereas the first $\alpha_{n-2} - r_{n-1}$ basis elements remain for the moment undesignated. The canonical form of the ($n-1$)-st bounding matrix is thus attained. We proceed in the same way up to step (1). Here the first $\alpha_0 - r_1$ canonical basis elements of C_0 do not remain undesignated but rather they are designated by $b_0^{j_0}$. This thus produces the canonical basis of $\mathfrak{C}$.

Now we can easily read off the homology groups of $\mathfrak{C}$ from the canonical bounding matrices. A chain d_q is obviously a cycle if and only if d_q in its representation by the canonical basis consists only of basis elements $b_q^{j_q}$ and $c_q^{i_q+1}$. In other words, Z_q is the submodule of C_q which is generated by the $b_q^{j_q}$ and $c_q^{i_q+1}$. A chain d_q is obviously a boundary if and only if d_q in its canonical basis representation is a linear combination of the $\partial a_{q+1}^{i_q+1} = e_{i_q+1}^{(q+1)} c_q^{i_q+1}$. In other words, B_q is the submodule of Z_q which is generated by these elements. It follows from this that, for the factor group $H_q = H_q(\mathfrak{C}) \approx Z_q/B_q$, H_q is the direct sum of the $\bar{\wp}_q$ infinite cyclic groups generated by the $b_q^{j_q}$ as representatives and the r_{q+1} finite cyclic groups of order $e_{i_q+1}^{(q+1)}$ which are generated by the $c_q^{i_q+1}$ as representatives.—Now it first of all follows that $\bar{\wp}_q = \wp_q$ is the q-dimensional Betti number. Our procedure yields the equations

$$\wp_q = \alpha_q - r_q - r_{q+1} \; (r_0 = r_{n+1} = 0), \tag{17.2}$$

which permit a simple determination of the Betti numbers of $\mathfrak{C}$ which uses only the ranks r_q of the bounding matrices.— Of the finite direct summands of $H_q(\mathfrak{C})$ of order $e_{i_q+1}^{(q+1)}$ naturally only those need to be taken into account for which this number is > 1. *The torsion numbers of $H_q(\mathfrak{C})$ are therefore the elementary divisors of the $(q + 1)$-st bounding matrix which are different from* 1. They also can be computed with the aid of the canonical bases in a finite number of steps. Moreover, H_n possesses no torsion numbers; H_n is a free group.

The elements of finite order of H_q form a finite subgroup of H_q which is called the *q-dimensional torsion group* $T_q = T_q(\mathfrak{C})$ of $\mathfrak{C}$. According to our results, T_q is the direct sum of the cyclic groups of order $e_{i_q+1}^{(q+1)}$ generated by the $c_q^{i_q+1}$. The factor group H_q/T_q is, according to our results, generated by the $b_q^{j_q}$ as representatives. It is a free group, therefore a module with the Betti number $\wp_q$ and is called the *reduced homology group* $\mathscr{H}_q = \mathscr{H}(\mathfrak{C})$ of $\mathfrak{C}$. It can, as one also sees from our results, be characterized as follows: let $\mathscr{B}_q = \mathscr{B}_q(\mathfrak{C})$ be the group of those cycles of $\mathfrak{C}$ a finite multiple of which is a boundary. We have that $B_q \subset \mathscr{B}_q$ and $\mathscr{B}_q$ is generated by precisely the r_{q+1} elements $c_q^{i_q+1}$ of the canonical basis. And H_q is represented as a direct sum, $H_q \approx \mathscr{H}_q + T_q$. The reduced homology groups $\mathscr{H}_q(\mathfrak{C})$ contain in fact less information about $\mathfrak{C}$ but they are, as free groups, simpler to handle than the full homology groups $H_q(\mathfrak{C})$, which we shall utilize in §9.

An important relation can be derived from the formulas (17.2). If we add these formulas for the various values of q with alternating sign $(-1)^q$, then all the ranks r_q cancel out and there results

$$\chi = \chi(\mathfrak{C}) = \sum_{q=0}^{n} (-1)^q \wp_q = \sum_{q=0}^{n} (-1)^q \alpha_q. \tag{17.3}$$

We call $\chi(\mathfrak{C})$ the *Euler characteristic* of $\mathfrak{C}$ and this equation is called the *Euler equation*. In the case that $\mathfrak{C}$ is the chain complex $\mathfrak{C}(\mathsf{K})$ of a simplicial complex K, its meaning lies in the fact that the Betti numbers $\wp_q$ (and hence also χ) are invariants of the polyhedron $R = |\mathsf{K}|$; χ can easily be read off from K as $\sum(-1)^q \alpha_q$. The simplest case of this formula is the *Euler polyhedra theorem* which, for triangulations of the 2-sphere where $\wp_0 = \wp_2 = 1$, $\wp_1 = 0$, therefore $\chi = 2$, asserts that

$$\alpha_0 - \alpha_1 + \alpha_2 = 2.$$

The Euler equation can be generalized, as we shall now show. Let $f: \mathfrak{C} \to \mathfrak{C}$ be a mapping of $\mathfrak{C}$ into itself. We note that f consists of boundary-preserving mappings of C_q into itself, $f_\bullet: C_q \to C_q$, which we now write as $f_\bullet = f_{\bullet q}$ to emphasize the dimension q for the present. To every such mapping $f_{\bullet q}$ there is assigned, according to the remark at the beginning of §16, a trace $\mathrm{Tr}(f_{\bullet q})$. Along with $f_{\bullet q}$ we consider the homomorphisms induced by $f_{\bullet q}$ of the reduced homology groups $\mathscr{H}_q(\mathfrak{C})$ (not of the usual homology groups $H_q(\mathfrak{C})$)

which we again write as $f_* = f_{*q}$ in order to emphasize the dimension q. Also the reduced homology groups $\mathscr{H}_q(\mathfrak{C})$ are modules whose mappings f_{*q} into themselves are assigned a trace $\mathrm{Tr}(f_{*q})$. The following important theorem is valid for these traces.

17.4 Theorem: (Lefschetz-Hopf Trace Formula): *If $f: \mathfrak{C} \to \mathfrak{C}$ is a mapping of the free chain complex $\mathfrak{C}$ into itself, then*

$$\sigma(f) = \sum_{q=0}^{n} (-1)^q \,\mathrm{Tr}(f_{\bullet q}) = \sum_{q=0}^{n} (-1)^q \,\mathrm{Tr}(f_{*q})$$

is valid. This number is called the trace invariant $\sigma(f)$ of f.

In particular, if $f = i$ is the identity mapping of $\mathfrak{C}$, then $f_{\bullet q}$ and f_{*q} are also identity mappings and obviously $\mathrm{Tr}(f_{\bullet q}) = \alpha_q$, $\mathrm{Tr}(f_{*q}) = \wp_q$ hold. Therefore $\sigma(i) = \chi(\mathfrak{C})$, the Euler characteristic of $\mathfrak{C}$. Theorem 17.4 thus contains the Euler equation as a special case.

Proof: $\mathrm{Tr}(f_{\bullet q})$ can be evaluated by means of every basis of C_q. We use here the canonical bases of Theorem 17.1 in which f can be expressed in the following form

$$\begin{aligned}
f_\bullet(a_q^{i_q}) &= \sum \alpha_{i_q \iota_q} a^{\iota_q} + \sum \alpha'_{i_q \eta_q} b_q^{\eta_q} + \sum \alpha''_{i_q \iota_{q+1}} c_q^{\iota_{q+1}},\\
f_\bullet(b_q^{j_q}) &= \qquad\qquad\quad \sum \beta_{j_q \eta_q} b_q^{\eta_q} + \sum \beta'_{j_q \iota_{q+1}} c_q^{\iota_{q+1}},\\
f_\bullet(c_q^{i_{q+1}}) &= \qquad\qquad\qquad\qquad\qquad\quad \sum \gamma_{i_{q+1} \iota_{q+1}} c_q^{\iota_{q+1}}.
\end{aligned}$$

In this connection, it must be remarked that all sums must be summed over the Greek indices; ι_q ranges over the same numbers as i_q, namely, $1, \ldots, r_q$; η_q ranges over the same numbers as j_q, namely $1, \ldots, \wp_q$. The fact that in the second row no α-sum and in the third row no α-sum and no β-sum occur follows from the fact that the image of a closed chain $b_q^{j_q}$ is again closed and that the image of a bounding chain $c_q^{i_{q+1}}$ is again a boundary. The traces mentioned in the theorem are now the numbers

$$\begin{aligned}
\mathrm{Tr}(f_{\bullet q}) &= \sum \alpha_{\iota_q \iota_q} + \sum \beta_{\eta_q \eta_q} + \sum \gamma_{\iota_{q+1} \iota_{q+1}},\\
\mathrm{Tr}(f_{*q}) &= \sum \beta_{\eta_q \eta_q};
\end{aligned}$$

the latter occurs because the $b_q^{j_q}$ form precisely a basis for the reduced homology groups $\mathscr{H}_q(\mathfrak{C})$. We make use now of the equation $f(\partial(a_q^{i_q})) = \partial(f(a_q^{i_q}))$. Upon using Theorem 17.4, it follows that

$$f_{\bullet}\partial a_q^{i_q} = f_{\bullet}(e_{i_q}^{(q)}\, c_{q-1}^{i_q}) = e_{i_q}^{(q)} \sum \gamma_{i_q \iota_q}\, c_{q-1}^{i_q},$$

$$\partial f_{\bullet} a_q^{i_q} = \partial\left(\sum \alpha_{i_q \iota_q} a_q^{\iota_q} + \sum \alpha'_{i_q \eta_q} b_q^{\eta_q} + \sum i''_{i_q \iota_{q+1}} c_q^{\iota_q+1}\right)$$

$$= \sum \alpha_{i_q \iota_q} e_{i_q}^{(q)} c_{q-1}^{\iota_q}.$$

Upon comparison of coefficients, it follows that

$$e_{i_q}^{(q)} \gamma_{i_q \iota_q} = \alpha_{i_q \iota_q} e_{\iota_q}^{(q)}.$$

In particular, for $i_q = \iota_q$ the $e_{i_q}^{(q)}$ cancel out, and there remains

$$\gamma_{\iota_q \iota_q} = \alpha_{\iota_q \iota_q}.$$

Therefore the trace of the matrix of the $\alpha_{i_q \iota_q}$ is equal to the trace of the matrix of the $\gamma_{i_q \iota_q}$. Hence, if we form the alternating sum of the $\mathrm{Tr}(f_{\bullet q})$, then these traces cancel each other out, entirely analogously as in the proof of the Euler equation the rank numbers r_q cancelled, and there results the alternating sum of the $\mathrm{Tr}(f_{*q})$. In this connection, one has to observe that for $q = n$ the $\gamma_{i_n \iota_n}$ and for $q = 0$ the $\alpha_{\iota_0 \iota_0}$ are missing. This completes the proof of the theorem.

We now go over to the investigation of the cohomology groups. As bases of the cochain groups C^q we take the bases dual to the canonical bases of the C_q, which are also called canonical bases. They likewise decompose into three types of cochains which we, with reference to Theorem 17.1 and according to our notational principles, have to name

$$a^q_{i_q},\ b^q_{j_q},\ c^q_{i_q+1}$$

for $i_q = 1, \ldots, r_q$ and $j_q = 1, \ldots, \wp_q$.

The canonical coboundary matrices are then transposed according to the introductory considerations of this section to those of the boundary matrices; we can visualize them by transposition of the matrices in Fig. 7. The equations read explicitly as follows:

$$\delta a^{q-1}_{i_{q-1}} = 0,\ \delta b^{q-1}_{j_{q-1}} = 0,\ \delta c^{q-1}_{i_q} = e^{(q)}_{i_q}\, a^q_{i_q}.$$

The structure of the cohomology groups can easily be realized from this. A cochain d^q is a cocycle if and only if it is a linear form in the $a^q_{i_q}$ and $b^q_{j_q}$ in its canonical representation; it is a coboundary if and only if it is a linear combination of the $e^{(q)}_{i_q}\, a^q_{i_q}$. Hence, H^q is the direct product of $\wp_q$ infinite cyclic groups which are generated by the $b^q_{j_q}$ as representatives, and of r_q finite cyclic groups of orders $e^{(q)}_{i_q}$ which are generated by the $a^q_{i_q}$ as representatives; of the last-named we naturally need only to take account of those for which $e^{(q)}_{i_q} > 1$. We have thus given a finite procedure for the determination of the cohomology groups.

Entirely analogously as in the homology groups we introduce here also the subgroups of all elements of finite order of H^q, the *cotorsion group* $T^q = T^q(\mathfrak{C})$. It is generated by the $a^q_{i_q}$ as representatives which have the given orders. The factor group H^q/T^q is isomorphic to the *reduced cohomology group* $\mathscr{H}^q = \mathscr{H}^q(\mathfrak{C})$, which is generated by the $b^q_{j_q}$ as representatives. And $\mathscr{H}^q$ can also be defined as the factor group of the cocycle group Z^q with respect to the group $\mathscr{B}^q$ of all those cochains of which a finite multiple is a coboundary. We have that H^q is the direct sum of $\mathscr{H}^q$ and T^q.—The comparison of our results for homology groups and cohomology groups now yields immediately the following theorem.

17.5 Theorem: *The homology groups H_q and the cohomology groups H^q of a free chain complex $\mathfrak{C}$ have the same Betti number $\wp_q$. They differ however in their torsion groups, we have that $T^q \approx T_{q-1}$.*

The last assertion results from the fact that T^q has the torsion number $e^{(q)}_{i_q}$, T_q has however the torsion numbers $e^{(q+1)}_{i_{q+1}}$. We have that $T_n \approx T^0 \approx 0$, H_n and H^0 are free groups. All of this is valid in particular in the case when $\mathfrak{C} = \mathfrak{C}(\mathsf{K})$ is the chain complex of a simplicial complex K. We then speak of the homology groups, reduced homology groups, torsion numbers, and so forth, of K, in symbols, $H_q(\mathsf{K})$, $\wp_q(\mathsf{K})$, and so on. In this case, something more special holds, that $H_0(\mathsf{K})$ possesses no torsion, i.e., that $T_0 \approx T^1 \approx 0$ and H_0 and H^1 are free groups (cf. Theorem 9.3).—The asymmetry which shows up here in the dimension indices touches upon the fact that the bounding matrices together with rank and elementary divisors are not properly assigned to a dimension q but correspond rather to the transition from the dimension q to the dimension $q - 1$; we could express this in symbols and thereby obtain symmetric, but more complicated, formulas.

For the canonical bases of the homology groups and the cohomology groups,

$$(a_q^{i_q}, a_{i_q}^q) = (b_q^{j_q}, b_{j_q}^q) = (c_q^{i_q+1}, c_{i_q+1}^q) = 1$$

hold as dual bases, while all the other Kronecker products of these basis elements vanish. Therefore

$$\Bigl(\sum_{j_q} h_{j_q} b_q^{j_q} + \sum_{i_q+1} k_{i_q+1} c_q^{i_q+1}, \sum_{i_q} g^{i_q} a_{i_q}^q + \sum_{j_q} h^{j_q} b_{j_q}^q\Bigr) = \sum h_{j_q} h^{j_q}$$

holds. Here, in the brackets for arbitrary integers h_{j_q}, k_{i_q+1}, g^{i_q}, h^{j_q} there appears on the left an arbitrary cycle z_q and on the right an arbitrary cocycle z^q. And therefore (z_q, z^q) is independent of the coordinates k_{i_q+1} of z_q and of the coordinates g^{i_q} of z^q. Whereas we know from Theorem 10.6 that (z_q, z^q) depends only on the homology class of z_q and the cohomology class and z^q, it turns out now that (z_q, z^q) depends only on the coordinates by h_{j_q} and h^{j_q} and hence only on the residue classes of z_q and z^q inside the *reduced* homology class and cohomology class. We can accordingly establish the following theorem.

17.6 Theorem: *The reduced homology groups $\mathscr{H}_q$ and cohomology groups $\mathscr{H}^q$ of the same dimension are mutually dual by virtue of the Kronecker product. To every q-dimensional homology basis of $\mathscr{H}_q$ there exists a dual q-dimensional cohomology basis of $\mathscr{H}^q$.*

PART III

CELL COMPLEXES. INVARIANCE

Chapter 5

CELL COMPLEXES

§ 18. Cell Decompositions

The simplexes which have served us as building blocks for polyhedra were in Chapter 4 also the foundation for the computation of the homology groups. Although the procedure developed there requires fundamentally only a finite number of steps, it is however relatively cumbersome in concrete cases. We can, for example, show that for the representation of a torus as a rectilinear simplicial complex at least fourteen triangles are necessary; the computation of the homology groups from the bounding matrices is formulated in a correspondingly complicated way. We shall now become acquainted with a method in which larger sets of simplexes of a simplicial complex K are combined into new entities, the so-called cells, which then form the foundation for a simpler computation of the homology groups. The cells are moreover important in our development for still other purposes, for example, for the invariance proof of the homology groups and for the theory of manifolds.

18.1 Definition: Let K be a simplicial complex. A q-dimensional subcomplex $\tau = \tau^q$ of K is called a *q-dimensional cell of* K, more specifically, a *q-dimensional homology cell,* if $H_q(\tau^q) \approx \mathsf{Z}$ (where Z is the group of integers) and $H_p(\tau^q) \approx 0$ for $p \neq q$.

We observe that $\mathrm{Bd}\tau = \mathrm{Hu}\tau - \tau$ is closed and that $H_q(\tau)$ can be described as the homology group of a subcomplex also as the homology group $H_q(\mathrm{Hu}\tau/\mathrm{Bd}\tau)$. We denote the boundary $\mathrm{Bd}\tau$ or briefly by $\dot{\tau}$ in the case of cells.—We have already denoted the system of homology groups $H_0 \approx 0, \ldots, H_{q-1} \approx H_q \approx \mathsf{Z}$ in §13 (after Theorem 13.3) as that of the homology groups of a q-dimensional cell.—A 0-dimensional cell τ^0 consists of a point; we have that $\mathrm{Hu}\tau^0 \approx \tau^0$, $\dot{\tau}^0 = \emptyset$. For $q \geqq 1$, we have that τ^q certainly is not a subcomplex of K by virtue of $H_0(\tau^0) \approx 0$.—*Examples* of cells: (1) A single q-dimensional simplex of a complex K is a q-dimensional cell.—(2) If K' is the normal subdivision of K then those simplexes of K' into which an open q-dimensional simplex of K decomposes forms a q-dimensional cell, as we shall see in §20.—(3) In the complex $\sum^2$ of the 2-dimensional sphere the star of a vertex is a 2-cell, as one easily verifies.—We shall become acquainted with further important examples later in the complementary cells of an n-dimensional manifold.

18.2 Definition: Let K be a simplicial complex. A set Π of cells of K is called a *cell decomposition* of K if the following two conditions are satisfied:

[CK 1] the cells form a decomposition of K;

[CK 2] the boundary $\dot{\tau}^q$ of every q-dimensional cell is the union set of certain cells τ_p with $p < q$. Π is also called a *cell complex.*

A simplicial complex K is itself trivially a cell complex with its simplexes as cells. Furthermore, the cells mentioned in the above example (2) form a cell decomposition of K' as we shall see.—If Π is a cell decomposition of K, then every simplex of K and also every point of $|\mathsf{K}|$ has, according to [CK 1], a uniquely determined carrier cell. The cells mentioned in [CK 2] are called the proper face cells of τ^q and written $\tau^p \prec \tau^q$; τ^q itself is as an improper face of τ^q. In the case of the trivial decomposition $\Pi = \mathsf{K}$ of K, this notion reduces to our earlier definition of the face simplexes and of the incidence $\sigma^p \prec \sigma^q$.—For a fixed dimension q, the set of all simplexes of K which are contained in any cell of dimension $p \leqq q$, is called the *q-dimensional skeleton* Π_q *of* Π. This definition is in agreement with our earlier definition of the skeleton K_q of K (see Volume I, §30). Observe that Π_q is not for instance a subset of Π but rather a subset of K. We have that $\Pi_q \subset K_q$ and $\Pi_n = K_n = K$, if K is n-dimensional.

18.3 Theorem: *Let $\sigma_1 \prec \sigma_2$ be two simplexes of K; let τ_1 and τ_2 be their respective carriers in a cell decomposition Π of K. Then either $\tau_1 = \tau_2$ or $\tau_1 \prec \tau_2$.*

Proof: We have that $\sigma_1 \in \mathrm{Hu}\sigma_2 \subset \mathrm{Hu}\tau_2$; therefore, either $\sigma_1 \in \tau_2$ or $\sigma_1 \in \dot{\tau}_2$. In the first case, $\tau_1 = \tau_2$. In the second case, it follows by [CK 2] that $\tau_1 \in \dot{\tau}_2$, i.e., $\tau_1 \prec \tau_2$.

18.4 Theorem: *The q-dimensional cell τ^q is open in Π_q. If $\tau^q < \tau^{q+1}$, then τ^q is open in $\dot{\tau}^{q+1}$.*

Proof: The boundary $\dot{\tau}^{q+1}$ of the $(q + 1)$-dimensional cell τ^{q+1} is a subcomplex of Π_q; therefore, the second assertion follows from the first.—It remains to show that if σ^p is a simplex of τ^q, further $\sigma^p \prec \sigma^r$ and $\sigma^r \in \Pi_q$, then it follows that $\sigma^r \in \tau^q$. In fact, it follows from $\sigma^p \prec \sigma^r$ according to the last theorem for the carrier cells $\tau(\sigma^p) = \tau^q$ of σ^p and $\tau(\sigma^r)$ of σ^r that either $\tau^q = \tau(\sigma^r)$ or $\tau^q \prec \tau(\sigma^r)$. But since $\sigma^r \in \Pi_q$ holds, we have that $\tau(\sigma^r)$ has at most the dimension q, so that the second possibility is excluded. Thus $\tau^q = \tau(\sigma^r)$ holds and that means $\sigma^r \in \tau^q$, which was to be proved.

We now proceed to show that to a cell decomposition Π there corresponds a chain complex $\mathfrak{C}(\Pi)$. To this end, we consider first a single cell $\tau = \tau^q$ of K. In this case, $H_q(\tau^q) \approx Z$ is equal to the group $Z_q(\tau^q)$, the group of the q-dimensional cycles of τ^q. This group has as an infinite cyclic group exactly two generators; they are called the *fundamental cycles* of τ^q. To orient the cell τ^q means to distinguish one of these two fundamental cycles as the positive d_p and the other as the negative $-d_q$. This corresponds thoroughly to the orientation of simplexes. We note the following theorem for later applications.

18.5 Theorem: *A q-dimensional cell $\tau = \tau^q$ determines uniquely up to sign a q-dimensional fundamental cycle d_q, which is the generator of the cyclic group $H_q(\tau^q)$.*

We now orient all the cells τ^q and choose a fixed fundamental cycle d_q for each τ^q. The chains d_q are at the same time chains of K in which the simplexes, except τ^q, occur with the coefficient zero. In every dimension q, all the d_q are linearly independent, since no two of them contain the same q-simplex with a coefficient $\neq 0$. They generate a subgroup D_q of $C_q(\mathsf{K})$ and are the free generators of it.

In order to investigate the bounding relations among the d_q we need a simple auxiliary consideration. Let N be an arbitrary subset of simplexes of a simplicial complex K and let c_q be a q-dimensional chain of K. By the "portion of c_q on N" we understand the sum of all simplexes of N perhaps with the coefficient with which the simplex occurs in c_q. We make the following assertion.

18.6 Theorem: *If* N *is open in* K, *then the portion on* N *of a cycle* c_q *of* K *is a cycle on* N. *The same holds when* c_q *is a chain of* K *whose boundary vanishes on* N.

It should be clear that the concept of a cycle in the assertion of the theorem is meant in the sense of the modified boundary formation ∂' in the subcomplex N.—We *prove* the second part of the theorem first. The assumption concerning c_q asserts the following. If s_{q-1} is an arbitrary elementary chain of N then ∂c_q has a coefficient zero for s_{q-1} in the representation as a sum of elementary chains. But at most those q-simplexes of c_q which are incident with s_{q-1} contribute to this coefficient. All these q-simplexes belong to N since N is open in K. Hence the equation $\partial' c_q = 0$ also holds in N, which was to be proved.

Now let d_q be the fundamental cycle of a cell τ^q. The boundary $\partial' d_q$ in the subcomplex τ^q vanishes and hence ∂d_q, the boundary in K, is a chain on $\dot{\tau}^q$. Let τ_j^{q-1} $(j = 1, 2, \ldots)$ range over the $(q-1)$-dimensional face cells of τ^q and let d_{q-1}^j be the fundamental cycle of τ_j^{q-1}. According to Theorem 18.4, τ_j^{q-1} is open in $\dot{\tau}^q$. Hence, according to the last theorem, the portion of ∂d_q on τ_j^{q-1} is a cycle on τ_j^{q-1}; therefore it is of the form $g_j d_{q-1}^j$ for some integer g_j. Since every $(q-1)$-simplex of τ^q occurs in one τ_j^{q-1}, ∂d_q is consequently of the form $\sum g_j d^j_{q-1}$, summed over j. Therefore ∂d_q belongs to D_{q-1}. It is thus shown that the boundary operator ∂ maps the group D_q into D_{q-1}. This means that the totality of groups D_q by virtue of ∂ form a chain complex $\mathfrak{D}$. But $\mathfrak{D}$ is a chain subcomplex of $\mathfrak{C}(\mathsf{K})$. As the chain complex $\mathfrak{C}(\mathsf{K})$ belongs to the simplicial complex K, the chain complex $\mathfrak{D}$ belongs to the cell complex Π. On the one hand, the elementary chains s_q correspond to the simplexes σ^q of K, where the s_q are free generators of the chain groups $C_q(\mathsf{K})$; on the other hand, the chains d_q correspond to the cells τ^q, where the d_q are free generators of the chain group D_q. In order to make this analogy still more evident, we introduce a new computational element t_q for each fundamental cycle d_q and hence for each cell τ^q of Π. We call t_q the *oriented cell of* τ^q, and we form with this t_q as generator the free abelian group which we call the *q-dimensional chain group* $C_q(\Pi)$ *of* Π.—By means of the one-to-one correspondence between the d_q and the t_q the D_q and $C_q(\Pi)$ are also in one-to-one correspondence. Further, for the t_q we introduce bounding relations which are fashioned after those for the d_q: Namely, if $\partial d_q = \sum g_j d^j_{q-1}$, then we define $\partial t_q = \sum g_j t^j_{q-1}$. Accordingly the groups $C_q(\Pi)$ form a chain complex, the *chain complex* $\mathfrak{C}(\Pi)$ *of the cell decomposition* Π. Here, $\mathfrak{C}(\Pi)$ and $\mathfrak{D}$ correspond one to one and are boundary preserving. We denote the transition from $\mathfrak{C}(\Pi)$ to $\mathfrak{C}(\mathsf{K})$ by ϑ: Let us set $\vartheta t_q = d_q$, from which there results by means of a linear extension the chain mapping

$$\vartheta\colon C_q(\Pi) \to C_q(\mathsf{K}).$$

Now ϑ maps $C_q(\Pi)$ into $C_q(\mathsf{K})$ and indeed onto $D_q \subset C_q(\mathsf{K})$; moreover, this mapping is boundary preserving, i.e., we have $\partial\vartheta = \vartheta\partial$. We call ϑ the *subdivision mapping* inasmuch as by means of ϑ the oriented cells t_q are decomposed in a certain way into the t_q combined simplexes of K. This relation will soon be still more obvious when we again encounter the mapping ϑ in connection with the normal decomposition of a simplicial complex. We summarize this in the following theorem.

18.7 Theorem: *A cell decomposition* Π *of a simplicial complex* K *determines a chain complex* $\mathfrak{C}(\Pi)$ *and accordingly for each dimension* q *a homology group* $H_q(\Pi)$.

In the next section we will show that $\mathfrak{C}(\Pi)$ and $\mathfrak{C}(K)$ have isomorphic homology groups and that ϑ_*, the homomorphism of $H_q(\Pi)$ into $H_q(K)$ induced by ϑ, represents an isomorphic mapping.

§ 19. The Homology Groups of Cell Decompositions

We retain the hypotheses and the relations of the preceding section.

19.1 Theorem: *Let* c_q *be a chain from* K *whose boundary* ∂c_q *lies in* D_{q-1}. *Then there exists a chain* b_q *from* D_q *for which* $b_q \sim c_q$, *i.e., for which* $\partial b_q = \partial c_q$ *and* $b_q - c_q \sim 0$ *hold.*

Proof: The proof splits up into three steps. (1) Let the dimension of K be n. Then $K = K_n = \Pi_n$, and c_q is a chain of Π_n. Let us assume that c_q is indeed a chain of Π_m with $n \geqq m > q$. We then construct a chain a_q from Π_{m-1} with $a_q \sim c_q$ as follows. Let τ^m be one of the m-dimensional cells, τ^m is open in Π_m. And c_q can be decomposed into $c_q = c_q{}^1 + c_q{}^2$, where $c_q{}^1$ is the portion of c_q on τ^m. And ∂c_q lies in D_{q-1}, therefore not in τ^m. Hence Theorem 18.6 is applicable, so that $c_q{}^1$ is a cycle on τ^m, that is, $\partial' c_q{}^1 = 0$. Since $H_q(\tau^m) = 0$, it follows further than $c_q{}^1 = \partial c_{q+1} + \bar{c}_q{}^1$ with suitable c_{q+1} from τ^m and for it a suitable $\bar{c}_q{}^1$ from $\dot{\tau}^m$, therefore from Π_{m-1}. Then

$$\bar{c}_q = \bar{c}_q{}^1 + c_q{}^2 = c_q{}^1 - \partial c_{q+1} + c_q{}^2 = c_q - \partial c_{q+1}$$

holds; furthermore, $\partial \bar{c}_q = \partial c_q$. It follows that $\bar{c}_q \sim c_q$. Now, $\bar{c}_q$ vanishes on τ^m, but outside of $\mathrm{Hu}\,\tau^m$ it remains unchanged with regard to c_q.—If we continue this procedure to all m-dimensional cells τ^m of Π, then we obtain one after another the chains c_q, $\bar{c}_q$, $\bar{\bar{c}}_q$, and so forth, and finally a chain a_q of the desired sort.—(2) By continuation of the procedure followed in (1), we can deduce c_q upon obtaining ∂c_q step by step to Π_{m-1}, Π_{m-2}, and so on, until we obtain a chain $b_q \in \Pi_q$ with $b_q \sim c_q$.—(3) We assert that then b_q possesses the property $b_q \in D_q$ required in the theorem. Namely, if τ_j^q ($j = 1, 2, \ldots$) is one of the q-cells of Π, and furthermore $b_q = b_q^1 + b_q^2$, where b_q^1 is the portion of b_q on τ_j^q, then b_q^1 is a cycle on τ_j^q as we see by the same deduction as was applied in (1). On account of $H_q(\tau_j^q) \approx Z_q(\tau_j^q) \approx \mathsf{Z}$, therefore b_q^1 is of the form $g_j d_q^j$, where d_q^j denotes the fundamental cycle of τ_j^q and g_j an integer. This holds for all j; therefore, b_q is of the form $\sum g_j d_q^j$, and, hence, belongs to D_q. This completes the proof of the theorem.

19.2 Theorem: (a) *Every q-dimensional cycle z_q of K is homologous to a q-dimensional cycle of D_q. (b) If a cycle y_q of D_q is the boundary of a chain of C_{q+1}, then it is also the boundary of a chain of D_{q+1}.*

Proof: The assertion (a) results from the preceding theorem if we apply it to z_q instead of to the chain c_q named there. The assertion (b) also follows when we set $y_q = \partial c_{q+1}$ and apply the preceding theorem to c_{q+1} instead of the chain c_q named there.

We now consider the inclusion mapping $i\colon \mathfrak{D} \to \mathfrak{C}(\mathsf{K})$ and the homomorphisms, induced by i in every dimension q, of the homology groups,

$$i_*\colon H_q(\mathfrak{D}) \to H_q(\mathsf{K}).$$

We assert that all the i_* are isomorphisms. In fact, part (a) of the last theorem asserts that i_* is epimorphic and part (b) that i_* is monomorphic.—This result can immediately be carried over from $\mathfrak{D}$ to the chain complex $\mathfrak{C}(\Pi)$ of the cell decomposition Π which stands in one-to-one correspondence with $\mathfrak{D}$ and of the homology groups $H_q(\Pi)$ of the latter. In place of i there then appears the subdivision mapping $\vartheta\colon C_q(\Pi) \to C_q(\mathsf{K})$. And ϑ induces accordingly in every dimension q an isomorphism

$$\vartheta_*\colon H_q(\Pi) \to H_q(\mathsf{K}).$$

This isomorphism is "natural" insofar as it depends only on K and Π.—We have thus attained the main result on cell decompositions, which we summarize in the next theorem.

19.3 Theorem: *The homology groups $H_q(\Pi)$ of a cell decomposition Π of a simplicial complex K are isomorphic to the homology groups $H_q(K)$ of K. The subdivision mapping ϑ induces a natural isomorphism of $H_q(\Pi)$ onto $H_q(K)$.*

With the aid of the cell decompositions one can compute in many cases the homology groups of a given polyhedron R without great difficulty. We sketch this in several cases, and, in particular, for several of the examples given in § 9, subsection 5.—We obtain a cell decomposition of the circular disk R by taking a point of the periphery as the cell τ^0, the remainder of the periphery as the cell τ^1 and the entire circular surface as the cell τ^2. Namely, we can take as the simplicial decomposition K of R a curvilinear triangle with its three edges and vertices, where one of the vertices is τ^0, the triangle as τ^2 and the subcomplex of the remaining edges and vertices as τ^1. With a suitable orientation of these cells, i.e., for a choice of suitable fundamental cycles t_q to these cells τ^q correspond the bounding relations: $\partial t_2 = t_1$, $\partial t_1 = 0$, from which the known homology groups of R result.—Also for an arbitrary simplicial decomposition K of the circular disk, we can do this with the corresponding three cells, namely with a vertex τ^0 on the boundary, the totality τ^2 of all interior simplexes and the subcomplex τ^0 of the remaining vertices and edges lying on the boundary. That τ^2 is really a cell, i.e., that the homology groups of τ^2 or the relative groups $H_q(\mathrm{Hu}\tau^2/\dot{\tau}^2)$ are really those of a cell, one can either compute directly or more simply deduce from the invariant character of the homology groups. As we shall prove in Chapter 6, the homology groups $H_q(K/\Lambda)$ depend only on the polyhedra $|K|$ and $|\Lambda|$, but not on the special simplicial decompositions K and Λ; if $|K|$ is a circular disk, $|\Lambda|$ its boundary, then the homology groups $H_q(K/\Lambda)$ are always those of a cell. —A cell decomposition of the 2-sphere is obtained by a point τ^0 and one two-dimensional cell τ^2; one-dimensional cells are not used at all. The bounding relations reduce to a single one: $\partial t_2 = 0$.—In an entirely analogous way, we can construct a cell decomposition of the n-sphere into a cell τ^0 and a cell τ^n with $\partial \tau_n = 0$, from which we read off directly the known homology groups of the sphere.—A cell decomposition of the torus R_1 is obtained as follows: A zero cell τ^0 consisting of a vertex p, two 1-cells $\tau_1{}^1$, $\tau_2{}^1$, namely a longitude circle and a meridian circle, which intersect in p, and a 2-cell τ^2. The bounding relations read $\partial t_2 = 0$, $\partial t_1{}^1 = \partial t$; $\partial t_1{}^2 = 0$, from which again the homology groups result directly.—We can look at the torus R_1 as a homeomorphic image of a sphere surface with one handle. We consider more generally the surface

R_g formed by a sphere surface with g handles, as indicated in Fig. 8 for $g = 2$. If we lay off on R_g, starting with an arbitrary initial point p of R_g, g generalized longitudinal circles and g generalized meridian circles

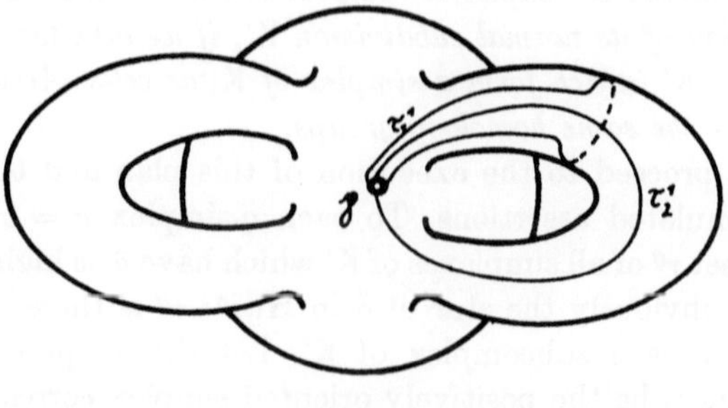

Fig. 8

τ^1_{1i}, τ^1_{2i} $(i = 1, \ldots, g)$, as indicated in the figure, then there results a cell decomposition of R_g into a 0-cell τ^0, $2g$ 1-cells as given, and a 2-cell τ^2. We should make it clear that τ^2 is a circular disk, therefore a cell. The bounding relations between the corresponding fundamental cycles read $\partial t_2 = 0$, $\partial t_1^{1i} = \partial t_1^{2i} = 0$. Hence R_g has the homology groups $H_2 \approx \mathsf{Z}$, $H_1 \approx 2g \times \mathsf{Z}$, $H_0 \approx \mathsf{Z}$; for the Betti numbers it follows that $\wp_2 = \wp_0 = 1$, $\wp_1 = 2g$.

§ 20. Normal Subdivisions

Let K be an n-dimensional complex, K′ its normal subdivision. Every open simplex (σ) of K decomposes—as was already emphasized in §4—into those simplexes of K′ which have the center of gravity $\hat{\sigma}$ as highest vertex. In this section, we shall show the following: For instance, if we combine in this way all simplexes of K′ into which a simplex (σ) decomposes, then we obtain a cell decomposition Π of K′. The incidences among the cells of Π are precisely the same as those among the corresponding simplexes σ of K. Hence we can, without confusion, identify Π with K, i.e., to regard K as a cell decomposition of K′. The result of the last section then asserts that K′ and the cell decomposition K of K′ possess the same homology groups. This indicates the invariance of the homology groups under normal subdivisions. We note these results as follows.

20.1 Theorem: *A simplicial complex* K *can be visualized as a cell decomposition of its normal subdivision* K′, *if we take for instance all simplexes of* K′, *which form a simplex of* K, *as cells. Accordingly,* K *and* K′ *have the same homology groups.*

We now proceed to the execution of this plan and to the proof of the formulated assertions. To each q-simplex $\sigma = \sigma^q$ of K we assign the set τ^q of all simplexes of K′ which have $\hat{\sigma}$ as highest vertex. Then τ^q is obviously the star of $\hat{\sigma}$ in $\mathrm{Hu}\tau^q$; τ^q is therefore open in $\mathrm{Hu}\tau^q$, i.e., it is a subcomplex of K′. Let the simplexes of K be oriented, let s_q be the positively oriented simplex corresponding to σ^q. We first introduce a mapping ϑ: $C_q(\mathsf{K}) \to C_q(\mathsf{K}')$ which will subsequently prove to be a subdivision mapping in the sense of §18. The definition materializes by complete induction on the dimension q by means of the following conventions:

$$(1)\,\vartheta s_0 = s_0, \quad (2)\,\vartheta s_q = \hat{s}_q \cdot \vartheta\partial s_q.$$

Here, s_0 denotes an arbitrary vertex of K which is also a vertex of K′. By (1) and by means of a linear extension, ϑ is defined on the entire group $C_0(\mathsf{K})$. And (2) is valid for $q = 1, 2, \ldots$ provided ϑ has already been defined on $C_{q-1}(\mathsf{K})$; $\hat{s}_q = \hat{\sigma}^q$ stands for the center of gravity of σ^q, the point in (2) designates the cone mapping in the interior of K′. By means of linear extension, then ϑ is defined on all of $C_q(\mathsf{K})$.— Now ϑ is boundary preserving, i.e., it satisfies the condition $\partial\vartheta = \vartheta\partial$, as we prove by induction: The assertion is trivial for zero chains s_0. Suppose it has already been proved for $(q-1)$-dimensional chains of K. Then for an elementary chain s_q we have that

$$\partial\vartheta s_q = \partial(\hat{s}_q \cdot \vartheta\partial s_q) = \vartheta\partial s_q - \hat{s}_q \cdot \partial(\vartheta\partial s_q)$$

according to the cone formula. In the last term, ∂s_q is of dimension $q - 1$; therefore we can apply the induction assumption and obtain $\partial\vartheta(\partial s_q) = \vartheta\partial\partial s_q = 0$ on account of $\partial^2 = 0$. From this the boundary preservation first follows for elementary chains s_q and then for the entire groups $C_q(\mathsf{K})$. We set $\vartheta s_q = d_q$.

We now prove by induction on the dimension n of K the following three assertions by which our theorem will then be proved.

(a_n) The τ^n are cells with the $\vartheta s_n = d_n$ as fundamental cycles.
(b_n) The totality of τ^q with $0 \leqq q \leqq n$ form a cell decomposition Π of K'; the incidence among the τ^q are the same as those among the corresponding σ^q of K.
(c_n) ϑ induces an isomorphism ϑ_*: $H_q(\mathsf{K}) \to H_q(\mathsf{K}')$.

For $n = 0$, all three assertions are trivial. The τ^0 are identical with the σ^0, and $\mathsf{K}' = \mathsf{K}$. Let the three assertions be proved for complexes K having dimension $<n$. We then first prove (a_n): τ^n was already recognized as a subcomplex of K' and as a star of $\hat{\sigma}$ in Huτ^n. Hence we can apply our earlier determination of the homology groups of stars (Theorem 13.5), which here yields

$$H_q(\tau^n) \approx H_{q-1}(\dot{\tau}^n) \text{ for } q = 2, \ldots, n,$$
$$H_1(\tau^n) \approx (k-1) \times \mathsf{Z},$$
$$H_0(\tau^n) \approx 0.$$

For $n = 1$, the number k of components of $\dot{\tau}^n$ is equal to 2, so that it follows that $H_1(\tau^1) \approx \mathsf{Z}$, $H_0(\tau^1) \approx 0$. The τ^1 are thus recognized as cells. For $n \geqq 2$ we have that $k = 1$; therefore, $H_0(\tau^n) \approx H_1(\tau^n) \approx 0$. And τ^n is a normal subdivision of the $(n-1)$-sphere $\sum^{n-1}$. According to the induction assumption (c_{n-1}), $H_{q-1}(\dot{\tau}^n) \approx H_{q-1}(\sum^{n-1})$; therefore, $H_{q-1}(\dot{\tau}^n) \approx \mathsf{Z}$ for $q - 1 = n - 1$, i.e., for $q = n$, and ≈ 0 for $q - 1 = 1, \ldots, n-2$, i.e., for $q = 2, \ldots, n-1$. These are therefore also the group types of $H_q(\tau^n)$. Thus, τ^n is also recognized as a cell for $n \geqq 2$. And $-d_n$ lies on $\dot{\tau}^n$, $\partial d_n = \partial\vartheta s_n = \vartheta\partial s_n$ lies on the cells to the boundary simplexes of σ^n, therefore *not* on τ^n. Therefore, d_n is a cycle on τ^n. Since d_n contains the n-simplexes of τ^n with coefficients ± 1, d_n is actually a fundamental cycle of τ^n.— To prove (b_n) we have only to establish that all τ^q of Π are cells; this results for those with $0 \leqq q < n$ according to the induction assumption (a_n), for those with $q = n$ according to what was just proved. Since the τ^q form a subdivision of K', they therefore really represent a cell decomposition Π of K'. The incidences among the t_q are, on account of $\partial d_q = \partial\vartheta s_q = \vartheta\partial s_q$, the same as those among the σ^q.—Finally, (c_n) is a simple consequence of (b_n). Because of the one-to-one and boundary preserving relation between Π and K we can, without fear of confusion, identify Π and K. The mapping ϑ: $C_q(\mathsf{K}) \to C_q(\mathsf{K}')$ is then precisely the subdivision mapping defined in §18 (there in the form ϑ: $C_q(\Pi) \to C_q(\mathsf{K})$) which according to the results of §19 induces the isomorphism ϑ_*: $H_q(\mathsf{K}) \to H_q(\mathsf{K}')$. This completes the proof of Theorem 20.1.

It is especially pointed out that we not only have proved the invariance of the homology groups under normal subdivision, but rather that we have also given a natural isomorphism, ϑ_*: $H_q(\mathsf{K}) \to H_q(\mathsf{K}')$, which depends only on K and K'. In this connection, the transition from K to K' transpires especially by the geometrically evident subdivision ϑ. We supplement this result further by giving also a simple geometric construction for the transition from K' to K, which induces the inverse isomorphism ϑ_*^{-1}: $H_q(\mathsf{K}') \to H_q(\mathsf{K})$. This occurs by means of a displacement η : $\mathsf{K}' \to \mathsf{K}$ (cf. §5, at the end). Namely, the following theorem is valid.

20.2 Theorem: *Let* K *be a simplicial complex and let* K' *be its normal subdivision, let* η: $\mathsf{K}' \to \mathsf{K}$ *be a displacement. The chain mapping* $\eta.$: $\mathfrak{C}(\mathsf{K}') \to \mathfrak{C}(\mathsf{K})$ *and the subdivision mapping* ϑ: $\mathfrak{C}(\mathsf{K}) \to \mathfrak{C}(\mathsf{K}')$ *stand in the relation* $\eta.\vartheta = 1$, *i.e., it is true for every chain* c_q *of K that* $\eta.\vartheta c_q = c_q$.

It suffices to prove this theorem for elementary chains s_q. It is trivial for $q = 0$ since a vertex of K is not changed by either ϑ or by $\eta.$. Assume the theorem has already been proved for chains of dimension $<q$. Let s_q be an arbitrary q-simplex of K. The vertices of all simplexes which appear in ϑs_q go over under $\eta.$, according to the definition of $\eta.$, into vertices of s_q. Therefore in all cases $\eta.\vartheta s_q = g s_q$ holds with an integer g. It follows that $\partial\eta.\vartheta s_q = \partial g s_q$, $\eta.\vartheta\partial s_q = g\partial s_q$. In the last equation, since ∂s_{q-1} has the dimension $n - 1$, we have, according to the induction hypothesis, that $g = 1$. Therefore $\eta.\vartheta s_q$

$= s_q$, which completes the proof of the theorem.—Moreover, the Sperner Lemma (Volume I, Theorem 34.1) is contained in this theorem, insofar as normal subdivisions are concerned. This theorem can moreover be generalized without difficulty to arbitrary subdivisions.—Now the next theorem follows directly from what has just been proven.

20.3 Theorem: *The displacement* η *and the subdivision mapping* ϑ *of the preceding theorem induce reciprocal isomorphisms* $\vartheta_*\colon H_q(\mathsf{K}) \to H_q(\mathsf{K}')$ *and* $\eta_*\colon H_q(\mathsf{K}') \to H_q(\mathsf{K})$. *Moreover* $\eta_*\vartheta_* = 1$.

We can also introduce the corresponding constructions for cochains and cohomology groups and prove the corresponding theorems. But we will not go into this in any detail here.

Chapter 6

INVARIANCE OF THE HOMOLOGY GROUPS

§ 21. Proof of Invariance

We now come to the proof of invariance for the homology groups. We assert: If R is a polyhedron and if K_1 and K_2 are two simplicial decompositions of R, then $H_q(\mathsf{K}_1) \approx H_q(\mathsf{K}_2)$ for every q. But we will actually show more: There exists a natural isomorphism between $H_q(\mathsf{K}_1)$ and $H_q(\mathsf{K}_2)$, to each homology class $h_1 \in H_q(\mathsf{K}_1)$ there is made to correspond in a natural way a well-defined equivalent homology class $h_2 \in H_q(\mathsf{K}_2)$ by virtue of the isomorphism. We first give a preliminary sketch how one can produce a connection between the two simplicial complexes K_1 and K_2, which of course in general have a very complicated disposition relative to one another. If $\mathsf{K}_1 < \mathsf{K}_2$, i.e., if K_1 and K_2 satisfy the star condition, this is easy provided one draws in a displacement $\eta \colon \mathsf{K}_1 \to \mathsf{K}_2$. If z_1 is a cycle of K_1 and $\eta . z_1 = z_2$ is the corresponding cycle of K_2, then, since η is homotopic to the identity mapping, we may look at z_1 and z_2 as being equivalent in a natural way. If K_1 and K_2 are arbitrary, then, by using a third simplicial decomposition K of R with $\mathsf{K} < \mathsf{K}_1$ and $\mathsf{K} < \mathsf{K}_2$, we can hope to produce a relation between K_1 and K_2 on the path through K.

We now proceed to carry out this sketch. In this connection we use a very simple auxiliary tool which is, however, especially convenient for the following and which we shall first develop. Suppose the indices ν, μ, λ, . . . range over an index set N, that to each of these indices ν there is a group G_ν and to each index pair ν, μ an isomorphism $\alpha(\nu\ \mu)$: $G_\mu \to G_\nu$ such that

$$\alpha(\nu, \mu) \circ \alpha(\mu, \lambda) = \alpha(\nu, \lambda) \text{ (transitivity relation).}$$

It then follows that $\alpha(\nu, \nu)$ is the identity isomorphism of G_ν and that $\alpha(\mu, \nu) = \alpha(\nu, \mu)^{-1}$. We call all these isomorphic groups which are related by the given isomorphisms a *transitive group system*. An element $g_\nu \in G_\nu$ and an element $g_\mu \in G_\mu$ can be called equivalent if $g_\nu = \alpha(\nu, \mu) g_\mu$. The totality of elements of all the groups G_ν thus decompose into equivalence classes. Every equivalence class contains exactly one element from each group G_ν, two equivalence classes are multiplied with one another by choosing for both factors representatives in the same group G_ν and multiplying these. The equivalence classes thus form the *model group* G_0 of the transitive group system. (The index zero is not yet contained in the index set N.) G_0 is isomorphic with all G_ν and there exist isomorphisms $\alpha(\nu, 0)$: $G_0 \to G_\nu$ and $\alpha(0, \nu)$: $G_\nu \to G_0$ such that the transitivity relations are still satisfied when the zero occurs among the participating indices.

We apply this to the set of all simplicial decompositions K_ν, $\nu \in \mathsf{N}$, of a polyhedron $R = |\mathsf{K}_\nu|$. We recall the relation $\mathsf{K}_\mu < \mathsf{K}_\nu$ of § 5 (at the end). It is satisfied, and thus there exists a homomorphism

$$\eta_* = \eta(\nu, \mu): H_q(\mathsf{K}_\mu) \to H_q(\mathsf{K}_\nu).$$

It results from a displacement η: $\mathsf{K}_\mu \to \mathsf{K}_\nu$ and is independent of the particular displacement used (Theorem 11.4). As follows directly from the form of the star condition in § 5, $\eta(\nu, \mu) \circ \eta(\mu, \lambda) = \eta(\nu, \lambda)$. If $\mathsf{K}_\mu = \mathsf{K}_\mu^{(m)}$ is the m-th normal subdivision of K_ν, then Theorem 20.3 shows that $\eta(\nu, \mu)$ is an *isomorphism*. We now show that this is true in general.

21.1 Theorem: *If* $\mathsf{K}_\mu < \mathsf{K}_\nu$, *then a displacement* η: $\mathsf{K}_\mu \to \mathsf{K}_\nu$ *always induces an isomorphism* $\eta_* = \eta(\nu, \mu)$ *of* $H_q(\mathsf{K}_\mu)$ *onto* $H_q(\mathsf{K}_\nu)$.

Proof: (a) $\eta(\nu, \mu)$ is epimorphic: We choose $\mathsf{K}_\lambda = \mathsf{K}_\nu^{(m)}$, the m-th normal subdivision of K_ν, so that $K_\lambda < K_\mu$, as is possible by Theorem 5.10. Then $\eta(\nu, \lambda)$ is an isomorphism. If $\eta(\nu, \mu)$ were not epimorphic, then obviously $\eta(\nu, \lambda) = \eta(\nu, \mu) \circ \eta(\mu, \lambda)$ could not be epimorphic.—(b) According to (a), also $\eta(\mu, \lambda)$ is epimorphic. Now if $\eta(\nu, \mu)$ were not monomorphic, then $\eta(\nu, \lambda) = \eta(\nu, \mu) \circ \eta(\mu, \lambda)$ could not be monomorphic.—The theorem follows from (a) and (b).

After this we write the isomorphism which is the inverse of $\eta(\nu, \mu)$ as $\eta(\nu, \mu)^{-1} = \eta(\mu, \nu)$. We now define the following isomorphism for arbitrary triangulations K_ν and K_μ of R: We choose, as is always possible, K_λ so that $K_\lambda < K_\mu$ and $K_\lambda < K_\nu$, and we set

$$\alpha(\nu, \mu) = (\eta(\nu, \lambda) \circ \eta(\lambda, \mu) \colon H_q(K_\mu) \to H_q(K_\nu).$$

This definition requires a justification: If also $K_{\lambda'} < K_\mu, K_\nu$, then we must show that by using λ' instead of λ the same isomorphism $\alpha(\nu, \mu)$ arises. Now if $K_{\kappa'} < K_\lambda, K_{\lambda'}$, then it follows that

$$\begin{aligned} \alpha(\nu, \mu) = \eta(\nu, \lambda) \circ \eta(\lambda, \mu) &= \eta(\nu, \lambda) \circ \eta(\lambda, \kappa) \circ \eta(\kappa, \lambda) \circ \eta(\lambda, \mu) \\ &= \eta(\nu, \kappa) \circ \eta(\kappa, \mu). \end{aligned}$$

We obtain the same result if λ' is used instead of λ. The definition of the isomorphism $\alpha(\nu, \mu)$ is thus unambiguous.

We now have to prove the transitivity of the $\alpha(\nu, \mu)$. For arbitrary K_ν, K_μ, K_λ we choose $K_\kappa < K_\nu, K_\mu, K_\lambda$. Then

$$\alpha(\nu, \mu) = \eta(\nu, \kappa) \circ \eta(\kappa, \mu), \ \alpha(\mu, \lambda) = \eta(\mu, \kappa) \circ \eta(\kappa, \lambda),$$

$$\alpha(\nu, \lambda) = \eta(\nu, \kappa) \circ \eta(\kappa, \lambda).$$

By a simple substitution, the assertion that $\alpha(\nu, \mu) \circ \alpha(\mu, \lambda) = \alpha(\nu, \lambda)$ is proven.—We summarize this result in the following theorem.

21.2 Theorem: *The homology groups $H_q(K_\nu)$ of the triangulations K_ν of a polyhedron R are mutually isomorphic, and by virtue of the isomorphism $\alpha(\nu, \mu)$, they form a transitive group system whose model group is called the homology group $H_q(R)$ of the polyhedron R.*

Because of the special significance of this theorem we explain the result once more. For every dimension q, there is a homology group

$H_q(R)$ assigned to the polyhedron R in an invariant manner. The homology group $H_q(\mathsf{K}_\nu)$ of an arbitrary triangulation K_ν of R is related to $H_q(R)$ through the isomorphism $\alpha(\nu, 0)$. If K_μ and K_ν are any two triangulations of R, then by means of $\alpha(\nu, \mu)$, there is uniquely assigned to each element of $H_q(\mathsf{K}_\mu)$ an equivalent element of $H_q(\mathsf{K})$. In particular, we can establish for each cycle of K_μ and K_ν, which may belong to entirely different triangulations, whether or not they represent equivalent homology classes, i.e., the same element of $H_q(R)$.—The investigations of the last paragraphs are hereby placed on a new foundation. The results on the homology groups of polyhedra have acquired an invariance significance.

§ 22. Supplements. Generalizations

We now again take up our consideration of transitive group systems. In addition to the system just considered of the groups G_ν and the isomorphisms $\alpha(\nu, \mu)$ with indices from the index set N, let the groups $G'_{\nu'}$ and isomorphisms $\alpha'(\nu', \mu')$, where the indices are taken from another index set N', be a second transitive group system. We say that a system of homomorphisms of the group G_ν into the groups $G'_{\nu'}$ is *compatible* if for each index pair ν, ν', there is a homomorphism $f(\nu', \nu)\colon G_\nu \to G'_{\nu'}$ and these homomorphisms are interrelated by the equation

$$f(\mu', \mu) = \alpha'(\mu', \nu')\circ f(\nu', \nu)\circ\alpha(\nu, \mu) \text{ (compatibility equation)}$$

for each two index pairs ν, ν' and μ, μ'. Such a compatible homomorphism system determines uniquely a model homomorphism

$$f(0', 0)\colon G_0 \to G'_{0'}$$

of the model groups, where $f(0', 0)$ is defined by $f(0', 0) = \alpha'(0', \nu')\circ f(\nu', \nu)\circ\alpha(\nu, 0)$.—And we see directly that $f(0', 0)$ is independent of the auxiliary index ν used.—We apply this to continuous mappings of polyhedra and their simplicial approximations.

Let $f\colon R \to R'$ be a continuous mapping of a polyhedron R into a polyhedron R'. Suppose K_ν ranges over all triangulations of R when

ν ranges over an index set N, and $K'_{\nu'}$ ranges correspondingly over all triangulations of R' when ν' ranges over the index set N'. In this section, $K_\nu \ll K'_{\nu'}$ means that K_ν and $K'_{\nu'}$ satisfy the star condition with respect to $f\colon |K_\nu| \to |K'_{\nu'}|$ (Definition 5.9). Then the following two assertions are valid.

(1) If $K_\mu < K_\nu \ll K'_{\nu'} < K'_{\mu'}$, then $K_\mu \ll K'_{\mu'}$.

(2) For given K_ν, $K'_{\nu'}$, there exists a $K_\mu < K_\nu$ with $K_\mu \ll K'_{\nu'}$.

By the definitions of the relations $<$ and $\ll$, (1) is clear, and (2) is a corollary of Theorem 5.10. Theorem 5.11 and Theorem 11.4 yield the following two results:

(3) If $K_\nu \ll K'_{\nu'}$, then there exist simplicial approximations $\varphi\colon K_\nu \to K'_{\nu'}$ of f, and each φ induces the same homomorphism $\varphi_* = \varphi(\nu', \nu)\colon H_q(K_\nu) \to H_q(K'_{\nu'})$.

(4) If $K_\mu < K_\nu \ll K'_{\nu'} < K'_{\mu'}$, then

$$\varphi(\mu', \mu) = \alpha'(\mu', \nu') \circ \varphi(\nu', \nu) \circ \alpha(\nu, \mu).$$

The latter follows from the form of the star condition. We can now make the following definition. If K_ν and $K'_{\nu'}$ are two arbitrary triangulations of R and R', respectively, and if K_μ is a triangulation of R as given in (2), let

$$f(\nu', \nu) = \varphi(\nu', \mu) \circ \alpha(\mu, \nu).$$

To justify this definition we must show that it is independent of the choice of μ. Therefore, let $K_\kappa < K_\nu$, $K_\kappa \ll K'_{\nu'}$. Choose $K_\lambda < K_\mu$, K_κ. Then

$$f(\nu', \nu) = \varphi(\nu', \mu) \circ \alpha(\mu, \nu) = \varphi(\nu', \mu) \circ \alpha(\mu, \lambda) \circ \alpha(\lambda, \mu) \circ \alpha(\mu, \nu)$$
$$= \varphi(\nu', \lambda) \circ \alpha(\lambda, \nu).$$

The same result would have been arrived at by considering κ instead of μ.

In order to show that the homomorphisms $f(\nu', \nu)$ form a compatible homomorphism system, the compatibility equations must be verified. But this is now easy: If $K_\kappa < K_\nu$, $K_\kappa \ll K'_{\nu'}$ and $K_\lambda < K_\mu$, $K_\lambda \ll K'_{\mu'}$, then choose K_ρ so that $K_\rho < K_\kappa$. We have $K_\rho < K_\lambda$. Then also $K_\rho < K_\nu$, $K_\rho \ll K'_{\nu'}$ and $K_\rho < K_\mu$, $K_\rho \ll K'_{\mu'}$ and we can set

$$f(\nu', \nu) = \varphi(\nu', \rho) \circ \alpha(\rho, \nu), f(\mu', \mu) = \varphi(\mu', \rho) \circ \alpha(\rho, \mu).$$

Hence,

$$f(u', \mu) = \varphi(\mu', \rho) \circ \alpha(\rho, \mu) = \alpha'(\mu', \nu') \circ \varphi(\nu', \rho) \circ \alpha(\rho, \nu) \circ \alpha(\nu, \mu)$$

$$= \alpha'(\mu', \nu') \circ f(\nu', \nu) \circ \alpha(\nu, \mu),$$

as was to be shown.—Thus, the $f(\nu', \nu)$ represent a compatible homomorphism system; they, therefore, determine a model homomorphism of the model group, which we denote by $f(0', 0)$.—We formulate the following theorem to supplement Theorem 21.2.

22.1 Theorem: *A continuous mapping f: $R \to R'$ of a polyhedron R into a polyhedron R' induces, for each q, a homomorphism*

$$f(0', 0) = f_*: H_q(R) \to H_q(R').$$

This is the model homomorphism of the compatible system of those homomorphisms which are derived from the totality of simplicial approximations of f.

The problem of invariance of the homology groups and the homomorphisms induced by them is solved in a satisfying manner by the last two theorems. For the cohomology groups of a polyhedron, one can show entirely analogous facts. We need not analyze here the considerations which run wholly parallel to those of §§ 21,22 and we leave it to the reader to carry out the details. The result is the following:

The cohomology groups $H^q(K_\nu)$ of the totality of triangulations K_ν of a polyhedron R are mutually isomorphic for every index q. They form a transitive group system by virtue of the well-defined isomorphisms $\beta(\mu, \nu)$: $H^q(K_\mu) \leftarrow H^q(K_\nu)$ between the homology groups of arbitrary triangulations K_μ, K_ν of R. These are defined as follows.

If $K_\lambda < K_\mu, K_\nu$, then the displacements of K_λ to K_μ and to K_ν induce isomorphisms:

$$\eta^*(\lambda, \nu): H^q(K_\lambda) \leftarrow H^q(K_\nu), \text{ and } \eta^*(\lambda, \mu): H^q(K_\lambda) \leftarrow H^q(K_\mu).$$

With $\eta^*(\mu, \lambda) = \eta^*(\lambda, \mu)^{-1}$, we set

$$\beta(\mu, \nu) = \eta^*(\mu, \lambda) \circ \eta^*(\lambda, \nu): H^q(K_\mu) \leftarrow H^q(K_\nu).$$

The model group of this transitive group system is defined to be the *q-dimensional cohomology group* $H^q(R)$ *of* R. To each continuous mapping $f: R \to S$ of the polyhedron R into the polyhedron S, there corresponds an *induced homomorphism*

$$f^*: H^q(R) \leftarrow H^q(S).$$

It is the model homomorphism for the compatible system of those homomorphisms which are derived from the totality of simplicial approximations of f.

Finally, we can immediately carry these results over to the relative homology groups and the relative cohomology groups. If $R = |K|$ is a polyhedron with the simplicial decomposition K, and Λ is a subcomplex of K, and $S = |\Lambda|$ is the corresponding subpolyhedron of R, then the homology groups $H_q(R/S)$ and the cohomology groups $H^q(R/S)$ are assigned to the pair R, S in an invariant manner. The homology groups and the cohomology groups of the totality of triangulations $(R, S) = (|K_\nu|, |\Lambda_\nu|)$ can be conceived of as a transitive group system with the aid of the isomorphisms $\alpha(\nu, \mu)$ and $\beta(\mu, \nu)$, respectively. By a continuous mapping $f: (R, S) \to (R' S')$, we mean a continuous mapping of the polyhedron R into the polyhedron R' which carries the subpolyhedron S of R into the subpolyhedron S' of R'. To each such mapping, there correspond the invariant homomorphisms

$$f_*: H_q(R/S) \to H_q(R'/S'),$$

$$f^*: H^q(R'/S') \to H^q(R/S).$$

Also the boundary operations ∂_* and δ^* of § 14 turn out to be invariant, but we cannot go further into this.

Regarding the relations we agree upon the following. We denote the homology classes of the polyhedron R, that is, the elements of the homology groups $H_q(R)$, by upper case Latin letters as Z_q, Y_q, and so on. If z_q is a cycle in any simplicial decomposition K of R, then we denote by Z_q the class of $H_q(R)$ determined by z_q. Analogously, we denote the cohomology classes of R by Z^q, Y^q, and so on, and the cohomology class of R which is determined by the cocycle z^q of a simplicial decomposition K of R by Z^q. If f is a continuous mapping of a polyhedron R into a polyhedron R', we denote the induced homomorphisms of the homology groups and cohomology groups by f_* and f^*. Thus, for instance, $f^*Z'^q = Z^q$ is the image of $Z'^q \in H^q(R')$ under f^*. By displacements η and subdivision mappings ϑ, the identity homomorphism is induced in the homology groups and cohomology groups, so that, for example, $\eta_* Z_q = Z_q$ or $\vartheta^* Y^q = Y^q$.

§ 23. Results and Applications

We now interrupt our development of homology theory in order to summarize and to make applications. In this connection, we look at the problems and examples as they were brought up in § 2 of the Introduction.

A. The Homeomorphism Problem

Let R and S be homeomorphic polyhedra and $f\colon R \to S$ be a homeomorphic mapping. By virtue of the mapping f which carries the topological structure of R in a one-to-one fashion into that of S, the construction of the homology group $H_q(R)$ determines step by step a corresponding construction of the homology group $H_q(S)$, and the two groups are thus isomorphic. The analogous fact holds for cohomology groups. We establish the following theorem.

23.1 Theorem: *Homeomorphic polyhedra have isomorphic homology groups and cohomology groups in every dimension. That is, a homeomorphism $f\colon R \to S$ determines in every dimension q an isomorphism* *between* $H_q(R)$ *and* $H_q(S)$ *and an isomorphism between* $H^q(R)$ *and* $H^q(S)$.

We give a second proof of this theorem, which uses, besides the homeomorphic mapping $f\colon R \to S$, also its inverse mapping $g = f^{-1}\colon S \to R$ and the induced homomorphisms

$$f_*\colon H_q(R) \to H_q(S),\ g_*\colon H_q(S) \to H_q(R).$$

It follows from $g \circ f = i_R$ that $g_* \circ f_*$ is the identity homomorphism of $H_q(R)$. If g_* were not monomorphic, then obviously also $g_* \circ f_*$ could not be monomorphic; if f_* were not epimorphic, then also $g_* \circ f_*$ could not be epimorphic. Hence g_* is monomorphic and f_* is epimorphic. On the basis of symmetry, also f_* is monomorphic and g_* is epimorphic. Thus, f_* is an isomorphism.—Moreover, $g_* = f_*^{-1}$. Similarly,

$$f^*\colon H^q(S) \to H^q(R) \text{ and } g^*\colon H^q(R) \to H^q(S)$$

are mutually inverse isomorphisms.

B. The Homotopy Problem

Let $f \simeq g\colon R \to S$ be two given homotopic mappings of two polyhedra R and S. According to Theorem 6.5, there exist simplicial approximations φ of f and ψ of g which belong to the same neighborhood class. And f and φ induce the same homomorphisms $f_* = \varphi_*$ of $H_q(R)$ into $H_q(S)$. We likewise have $g_* = \psi_*$. Finally, according to Theorem 11.2, φ and ψ induce the same homomorphisms $\varphi_* = \psi_*$. From this it follows that $f_* = g_*$. Utilizing the cohomology groups, we formulate the following theorem.

23.2 Theorem: *Homotopic mappings* $f \simeq g\colon R \to S$ *induce in every dimension* q *the same homomorphisms*

$$f_* = g_*\colon H_q(R) \to H_q(S) \textit{ and } f^* = g^*\colon H^q(S) \to H^q(R).$$

In the example (c) in § 2, *B*, we have that $H_n(S^n) \approx \mathbf{Z}$. In the n-th dimension, f_* is the identity automorphism of $H_n(S^n)$; in contrast we easily see that g_* carries a generator α of $H_n(S^n)$ into

$(-1)^{n-1}\alpha$. Therefore, for even n, g_* is different from f_* so that it follows, in this case, that $f \not\simeq g$.—In example (d), f_* is again the identity automorphism. On the other hand, g_* is the zero homomorphism for $q > 0$, so that $f \not\simeq g$.—We give another example for two spaces R and S and the totality of classes of mappings f: $R \to S$: Let $R = S = S^1$ be the circumference of the unit circle, described in the usual way in R by means of the argument angle $\varphi \bmod 2\pi$ and in S by means of the argument angle $\psi \bmod 2\pi$. We assert: The mappings $\psi = f_k(\varphi) = k\varphi$ for $k = 0, \pm 1, \ldots$ represent all mapping classes and each one exactly once. That these $f_k(\varphi)$ belong to distinct classes follows from the fact that f_{k*} carries the generator α of $H_q(R)$ over into the k-fold, $k\beta$, of the generator β of $H_q(S)$. That every mapping f: $R \to S$ is homotopic to one of the mappings $f_k(\varphi)$ can likewise be proved without too much difficulty; we do not carry this out in detail here.

C. Special Homotopies

1. If f: $R \to S$ is an unessential mapping of two polyhedra, i.e., $f \simeq c$ with a constant mapping c, then the homomorphisms c_*: $H_q(R) \to H_q(S)$ in the same dimensions $q > 0$ are zero homomorphisms, and hence the same is valid for the homomorphisms f_*. Therefore, *a necessary condition for the inessentiality of f is that the homomorphisms f_* in the dimensions $q > 0$ are zero homomorphisms.* For the identity mapping i of the n-sphere for $n \geqq 1$, this is not the case; i is, therefore, essential.

2. For a deformation of the identity f: $R \to R$, we have that $f \simeq i_R$. It follows that *if f is a deformation of the identity, then all induced homomorphisms f_*: $H_q(R) \to H_q(R)$ are the identity homomorphisms.* From 1. and 2. taken together, it follows that:

3. *Necessary conditions for a space R to be contractible are that $H_q(R) \approx 0$ holds for all $q > 0$.*

4. If the polyhedron A is a retract of the polyhedron R (§ 2, C 4) and if r: $R \to A$ is the corresponding retracting mapping and i: $A \to R$ the inclusion mapping, then we have that $r \circ i = i_A$. It

follows that $r_* \circ i_*$ is the identity homomorphism of $H_q(A)$ onto itself. In order to evaluate this fact, we need several simple facts about abelian groups which we now insert here.

23.3 Theorem: *If f: $G \to F$ and g: $F \to G$ are homomorphic mappings of the abelian groups G, F whose composition $g \circ f = i_G$: $G \to G$ is the identity homomorphism of G, then f is monomorphic and g is epimorphic; furthermore, F possesses the representation $F = \operatorname{Im} f \dotplus \operatorname{Ker} g$ as a direct sum, where $\operatorname{Im} f$ is isomorphic to G.*

In this connection, $\operatorname{Im} f$ denotes as usual the *image* of G under f and is a subgroup of F; further, $\operatorname{Ker} g$ denotes the *kernel* of g, namely the subgroup of all elements x of F with $g(x) = 0$.

Proof: If x_1 and x_2 are elements of G with $f(x_1) = f(x_2)$, then it follows that $g(f(x_1)) = g(f(x_2))$ and, since $g \circ f = i$, we have that $x_1 = x_2$. Therefore, f is monomorphic.—For every $x \in G$, we have that $x = g(f(x))$; x is, therefore, an image under g; thus, g is epimorphic. (We have already applied this deduction method in the second proof of Theorem 23.1.)— If $y \in F$ belongs to $\operatorname{Im} f$, i.e., $y = f(x)$ for a suitable x, and y also belongs to $\operatorname{Ker} g$, i.e., $g(y) = 0$, then it follows that $0 = g(y) = g(f(x)) = x$; hence, $y = 0$. Therefore, $\operatorname{Im} f$ and $\operatorname{Ker} g$ have intersection 0. It remains to show that every $y \in F$ possesses a representation $y = y_1 + y_2$ with $y_1 \in \operatorname{Im} f$ and $y_2 \in \operatorname{Ker} g$. To this end, we set $y_1 = f(g(y))$ and $y_2 = y - y_1$, so that $y = y_1 + y_2$. It is clear that $y_1 \in \operatorname{Im} f$. Further, $g(y_2) = g(y) - g(y_1) = g(y) - g(f(g(y))) = g(y) - g(y) = 0$, i.e., $y_2 \in \operatorname{Ker} g$, which was to be shown.—This immediately implies the following theorem.

23.4 Theorem: *If f: $G \to F$ and g: $F \to G$ are homomorphic images of the groups G and F whose products $g \circ f = i_G$ and $f \circ g = i_F$ are the identity homomorphisms of G and F, respectively, then G and F are isomorphic and f and g are inverse isomorphisms of one another.*

We return now to retracts. Theorem 23.3 is directly applicable to the homomorphisms i_*: $H_q(A) \to H_q(R)$ and r_*: $H_q(R) \to H_q(A)$. From this there results the following theorem.

23.5 Theorem: *If the polyhedron A is a retract of the polyhedron R, then the homomorphisms i_* induced by the inclusion mapping i: $A \to R$ are monomorphic and the homomorphisms r_* induced by the retracting mapping r: $R \to A$ are epimorphic. Furthermore, $H_q(R)$ possesses a direct summand which is isomorphic to $H_q(A)$.*

The theorem gives an easily verifiable necessary criterion for a subpolyhedron A of the polyhedron R to be a retract of R. We can, for example, say: *The n-dimensional solid sphere $R = T^n$ does not have its $(n-1)$-dimensional boundary sphere $S = S^{n-1}$ as a retract*; for, $H_{n-1}(T^n) \approx 0$ does not possess a direct summand which is isomorphic to $H_{n-1}(S^{n-1}) \approx \mathsf{Z}$. We elucidate the meaning of this theorem by applying it to prove the following important theorem.

23.6 Theorem: (Brouwer Fixed Point Theorem): *A continuous mapping f of the n-dimensional solid sphere T^n into itself, or of the closed n-simplex $[\sigma^n]$ into itself, possesses at least one fixed point, i.e., a point p with $f(p) = p$.*

Proof: Let T^n be the n-dimensional unit solid sphere in Euclidean n-space R^n. Let us assume that f has no fixed point. Then, for every point $x \in T^n$, the vector from $f(x)$ to x is well-defined; let its point of intersection with the boundary sphere S^{n-1} be $g(x)$. Obviously, $g(x)$ is a continuous mapping of T^n into itself. Further, $g(x) = x$ for all $x \in S^{n-1}$; therefore, $g(x)$ is a retraction of T^n onto S^{n-1}. Thus, S^{n-1} turns out to be a retract of T^n. This is in contradiction to our above result, and hence the theorem is proved.

5. If the polyhedron A is a deformation retract of the polyhedron R (§ 2, C 5), then besides $r \circ i = i_A$, we have $i \circ r \simeq i_R$. For the induced homomorphisms, it follows that $r_* \circ i_*$ as well as $i_* \circ r_*$ is the identity automorphism. This, by virtue of Theorem 23.4, implies the following theorem.

23.7 Theorem: *A polyhedral deformation retract A of the polyhedron R has the same homology groups as R itself.*

Hence, for example, a single point p of the 1-sphere S^1 represents a retract of S^1, but not a deformation retract of S^1, as we have already asserted in § 2.—Up to this point we have derived under *C* only criteria concerning homology groups; we can supplement it immediately by criteria concerning cohomology groups. We do not go further into this.

D. Homotopy Types

If R and S are two polyhedra of the same homotopy type in the sense of Definition 2.9, then, in the notation of this definition, the homomorphisms $g_* \circ f_*$ of $H_q(R)$ onto itself and the homomorphisms $f_* \circ g_*$ of $H_q(S)$ onto itself are the identity homomorphisms. Hence it follows, again by Theorem 23.4, that $H_q(R) \approx H_q(S)$. We can thus state the following theorem.

23.8 Theorem: *Polyhedra of the same homotopy type have isomorphic homology groups and cohomology groups.*

Accordingly, the homology groups and the cohomology groups are not suitable for distinguishing different space types of the same homotopy type from one another, which fact we have already pointed out in § 2.

E. Fixed Point Theorems

As a further application of homology theory, we investigate the fixed points of a mapping f of a polyhedron R into itself, that is, the points $\mathfrak{x} \in R$ with $f(\mathfrak{x}) = \mathfrak{x}$. They are interrelated with the so-called trace invariant $\varphi(f)$ of f, which is defined as follows: f induces in the reduced homology groups $\mathscr{H}_q(R)$ the homomorphisms f_{*q}: $\mathscr{H}_q(R) \to \mathscr{H}_q(R)$, and since the $\mathscr{H}_q(R)$ are modules, there corresponds to each of these homomorphisms, a trace $\mathrm{Tr}(f_{*q})$. Then $\sigma(f)$ is defined by

$$\sigma(f) = \sum_{q=0}^{n} (-1)^q \, \mathrm{Tr}(f_{*q}).$$

This agrees with the definition given in § 17 of the trace of a mapping of a chain complex into itself. Also here we will represent $\sigma(f)$ as the trace of a suitable chain mapping. We first formulate the following theorem.

23.9 Theorem: *For a fixed-point-free mapping $f: R \to R$ of a polyhedron R into itself, the trace invariant $\sigma(f)$ of f vanishes.*

Proof: Since f is fixed-point-free, it is true for all $\mathfrak{x} \in R$ that $f(\mathfrak{x}) \neq \mathfrak{x}$. Since R is compact, there exists an $\epsilon > 0$ such that $d(\mathfrak{x}, f(\mathfrak{x})) > \epsilon$ for all $\mathfrak{x} \in R$. We choose a simplicial decomposition K of R whose simplex diameters are all smaller than $\epsilon/2$ and then a subdivision $\mathsf{K}' = \mathsf{K}^{(\nu)}$ of K such that f has a simplicial approximation $\varphi: \mathsf{K}' \to \mathsf{K}$. In addition to the chain mapping $\varphi_{\bullet q}: C_q(\mathsf{K}') \to C_q(\mathsf{K})$ corresponding to φ, we also consider the subdivision mapping $\vartheta_q: C_q(\mathsf{K}) \to C_q(\mathsf{K}')$ and the mapping $\psi_q = \vartheta_q \varphi_{\bullet q}: C_q(\mathsf{K}') \to C_q(\mathsf{K}')$. We can represent $\mathscr{H}_q(R)$ in the sense of § 21 as we please by $\mathscr{H}_q(\mathsf{K})$ or $\mathscr{H}_q(\mathsf{K}')$. Both these groups are interrelated by the isomorphism $\vartheta_{*q}: \mathscr{H}_q(\mathsf{K}) \to \mathscr{H}_q(\mathsf{K}')$ (Theorem 20.3); we choose the second possibility. Similarly, f_* can be represented by $\varphi_*: \mathscr{H}_q(\mathsf{K}') \to \mathscr{H}_q(\mathsf{K})$ or also by $\psi_* = \vartheta_* \varphi_*: \mathscr{H}_q(\mathsf{K}') \to \mathscr{H}_q(\mathsf{K}')$; we choose the second possibility here also. According to the Lefschetz-Hopf trace formula (Theorem 17.4), we then have to prove that

$$\begin{aligned}\sigma(f) &= \sum(-1)^q \operatorname{Tr}(f_{*q}) \\ &= \sum(-1)^q \operatorname{Tr}(\psi_{*q}) = \sum(-1)^q \operatorname{Tr}(\psi_{\bullet q}) = 0.\end{aligned}$$

Two points whose closed carriers in K contain common points have, under our assumptions, a distance $< \epsilon$. Since $d(\mathfrak{x}, f(\mathfrak{x})) > \epsilon$ for all $\mathfrak{x} \in R$, the closed carriers in K of $\mathfrak{x}$ and $f(\mathfrak{x})$ are disjoint. We can say exactly the same thing concerning the closed carriers in K of $\mathfrak{x}$ and $\varphi(\mathfrak{x})$, since $\varphi(\mathfrak{x})$ lies on the closed carrier of $f(\mathfrak{x})$. Hence, if s_q' is an oriented simplex of K' with the carrier simplex σ_p in K, then the image $\varphi_{\bullet q} s_q'$ is either 0 or an oriented q-simplex of K, which has no points in common with $[\sigma^p]$. Therefore, s_q' does not occur in $\psi_q s_q' = \vartheta_q \varphi_{\bullet q} s_q'$ as a summand. But this means that $\operatorname{Tr}(\psi_{\bullet q}) = 0$. This completes the proof of the theorem.

We elucidate the significance of the theorem just proved by

several further remarks.—The trace invariant $\sigma(f)$ is determined by the homomorphism f_* induced by f of the reduced homology groups, and, hence, it depends only on the homology behavior of f. Accordingly, in particular, the homotopic mappings of R into itself have the same trace invariant. All statements which we can make on the basis of the trace invariant $\sigma(f)$ concerning fixed points pertain to f and also to all mappings of the mapping class of f.—$\sigma(f) = 0$ is necessary for f to be fixed-point-free, but it is in no way sufficient, as we can immediately see from the simplest examples. By what we just said, $\sigma(f) = 0$ is also necessary for the existence of a fixed-point-free mapping which is homotopic to f. Also for this, $\sigma(f) = 0$ is not sufficient.—By the theorem just proved, the fact that $\sigma(f) = 0$ can be made useful for statements concerning fixed points of f. The same is also possible for $\sigma(f)$ itself: $\sigma(f)$ measures in a certain way the number of fixed points of f. But we cannot go into this detail here.

In the application of Theorem 23.9 it is to be noted that for connected polyhedra R, we always have that $\mathrm{Tr}(f_{*0}) = 1$, since a positively oriented vertex of $R = |\mathsf{K}|$ always goes over as the basis of the 0-dimensional homology group into a positively oriented vertex.— Now let R be the n-dimensional solid sphere or, what amounts to the same thing, a closed n-dimensional simplex and let f be an arbitrary mapping of R into itself. Since $\wp_q = 0$ for all $q > 0$, also all $\mathrm{Tr}(f_{*q})$ vanish for $q > 0$; it follows that $\sigma(f) = \sigma(f_{*0}) = 1$. By the last theorem, f has at least one fixed point. The *Brouwer fixed point theorem* is thus once more proved for the n-dimensional solid sphere (Theorem 23.6).

Now let $R = S^n$ be the n-dimensional sphere and let f be an arbitrary mapping of S^n into itself. We have that $\mathrm{Tr}(f_{*0}) = 1$, and since $\wp_q = 0$ for $0 < q < n$, it follows that $\mathrm{Tr}(f_{*q}) = 0$ for these q. And $\mathrm{Tr}(f_{*n})$ is nothing else than the mapping degree γ of f, as follows directly from the definition of γ. Thus, it follows that $\sigma(f) = 1 + (-1)^n\gamma$. Therefore, the trace invariant $\sigma(f)$ is determined solely by γ. In particular, if f is fixed-point-free, then it follows that $\sigma(f) = 0$.

23.10 Theorem: *A mapping f of the n-sphere into itself without fixed points has the mapping degree $\gamma = (-1)^{n-1}$.*

Let us now assume that R is an arbitrary polyhedron and that f is the identity mapping or a deformation of the identity mapping, or as we shall say briefly a deformation. Then obviously $\mathrm{Tr}(f_{*q}) = \wp_q$, and the next theorem follows.

23.11 Theorem: *The trace invariant $\sigma(f)$ of a deformation of a polyhedron R is equal to the Euler characteristic of R:*

$$\sigma(f) = \chi(R) = \sum_{q=0}^{n} (-1)^q \wp_q.$$

From this theorem there results a simply verifiable criterion for a deformation f of a polyhedron R to have fixed points. We consider, as an example, the surfaces F mentioned in the fundamental theorem of surface theory (immediately following Theorem 27.2) and we investigate for which of them the necessary condition of Theorem 23.9 is satisfied. For orientable surfaces F, $\chi(F) = 2 - 2g$ holds, and it follows from $\chi(F) = \sigma(F) = 0$ that $g = 1$, whereby the torus is characterized. For non-orientable surfaces F, $\chi(F) = 2 - k$ holds, and it follows from $\chi(F) = \sigma(F) = 0$ that $k = 2$, whereby the Klein bottle is characterized. Since these two surfaces obviously admit fixed-point-free deformations, we can summarize this in the following theorem.

23.12 Theorem: *Among the connected 2-dimensional manifolds only the torus and the Klein bottle allow fixed-point-free deformations.*

Finally, we apply these theorems to deformations of spheres. The following important theorem is valid.

23.13 Theorem: *The n-sphere S^n allows a fixed-point-free deformation if and only if n is odd.*

Proof: A deformation of the identity mapping has degree 1. If it possesses no fixed points, then, according to Theorem 23.10, it has the degree $(-1)^{n-1}$; therefore, in this case, n must be odd. This completes the proof of half of the theorem.—The existence of a fixed-point-free deformation for spheres S^n of odd dimension $n = 2m - 1$ can easily be seen. Let the $2m$-dimensional Euclidean space R^{2m} be described by m complex coordinates $z_1, \ldots, z_m$, the n-dimensional unit sphere in it by the equation $z_1\bar{z}_1 + \ldots + z_m\bar{z}_m = 1$, where the bars denote the complex conjugates. Then, for every $0 \leqq t \leqq 1$, let a mapping f_t of S^n into itself be defined by $f_t(z_j) = z_j e^{it}$. Then $f = f_1$ is obviously a fixed-point-free mapping of S^n, and f_t represents a deformation of f into the identity.—We deduce from this theorem a significant consequence concerning continuous vector fields on spheres.

23.14 Theorem: *An n-sphere S^n possesses a nowhere vanishing tangential continuous vector field if and only if n is odd.*

In particular, there is no nowhere vanishing tangential continuous vector field on the 2-sphere (a porcupine just cannot be combed).

Proof: Let $\mathfrak{v}$ be a nowhere vanishing tangential continuous vector field on the n-sphere S^n which is realized as the unit sphere with center at O in the Euclidean space R^{n+1}. Let $\mathfrak{v}(\mathfrak{x})$ be the vector piercing in the point $\mathfrak{x}$ of S^n, let $\mathfrak{y}_t$ with $0 \leqq t \leqq 1$ be the endpoint of the vector $t\mathfrak{v}(\mathfrak{x})$. Let the point at which the joining segment $O\mathfrak{y}_t$ touches the sphere S^n be denoted by $f_t(\mathfrak{x})$. Then $f = f(\mathfrak{x}) = f_t(\mathfrak{x})$ is a continuous mapping of S^n into itself without fixed points and $f_t(\mathfrak{x})$ is a homotopy joining f with the identity mapping $f_0(\mathfrak{x})$ of S^n. According to the preceding theorem n must thus be odd. This proves half of the theorem.—The existence of a nowhere vanishing tangential continuous vector field on a sphere S^n of odd dimension $n = 2m - 1$ can easily be seen. Let the $2m$-dimensional Euclidean space R^{2m} be described by means of real cartesian coordinates $(x_1, \ldots, x_{2m})$ and let S^n be represented as the $(2m - 1)$-dimensional unit sphere in R^{2m}. If $\mathfrak{x} = (x_1, \ldots, x_{2m})$ is a point of S^n, then let the vector $\mathfrak{v}(\mathfrak{x})$ with the coordinates $(-x_{m+1}, \ldots, -x_{2m}, x_1, \ldots, x_m)$

be assigned to it. The vector field $\mathfrak{v} = \mathfrak{v}(\mathfrak{x})$ obviously has the desired properties.

§ 24. Local Homology Groups

Let R be a polyhedron, $\mathfrak{p}$ a point of R. We will introduce a new sort of homology groups which depend only on $\mathfrak{p}$ and on a neighborhood of $\mathfrak{p}$ in R which can be chosen arbitrarily small. We call them local homology groups $H_q(R, \mathfrak{p})$ of R at the point $\mathfrak{p}$. These groups belong to the so-called "properties in the small" of the space R, whereas the ordinary homology groups $H_q(R)$ as "properties in the large" depend on the polyhedron R in its entirety. In order to arrive at an invariant definition of the local homology groups, we proceed by a method similar to the one used in the definition of the groups $H_q(R)$. There we determined the homology groups first for a special simplicial decomposition K of R. Then we considered the totality of simplicial decompositions K_ν of R—ν from a suitable index set—and showed that all these K_ν lead, in every dimension q, to isomorphic groups $H_q(\mathsf{K})$. The common group type so determined was recognized as an invariant of the polyhedron R and was used as an exact definition of $H_q(R)$. We now consider closed neighborhoods of $\mathfrak{p}$ which admit a simplicial decomposition in a closed star $\mathsf{H} = \mathrm{Hu}\,\Theta$ with $\Theta = \mathrm{St}\,\mathfrak{p}$ with the midpoint $\mathfrak{p}$. Here, H need not be the star of $\mathfrak{p}$ in a simplicial decomposition of all of K. We will later show: (1) that arbitrarily small stars of this type exist, i.e., that such stars exist in every neighborhood of $\mathfrak{p}$, and (2) that all these stars about $\mathfrak{p}$ lead, in every dimension q, to isomorphic homology groups $H_q(\Theta)$. In this, we mean the homology groups $H_q(\Theta)$ introduced in Theorem 13.3 on the basis of the modified boundary formation ∂' in the subcomplex Θ, i.e., if we set $\mathrm{Bd}\,\Theta = \mathsf{H} - \Theta = \mathsf{P}$, then $H_q(\Theta) = H_q(\mathsf{H}/\mathsf{P})$ is the corresponding relative homology group of H with respect to P. It follows from (2) that the group type of $H_q(\Theta)$, which is the same for all stars Θ about $\mathfrak{p}$, is an invariant of the polyhedron R with respect to $\mathfrak{p}$ and hence can be used as a definition of $H_q(R, \mathfrak{p})$. Moreover, it follows from (1) that $H_q(R, \mathfrak{p})$ really depends only on an arbitrarily small neighborhood of $\mathfrak{p}$ in R.—We postpone the proof of both assertions (1) and (2) till the end of this section and we assume them foı the moment as known. With this assumption, we define the *q-dimensional local homology group* $H_q(R, \mathfrak{p})$ *of* R *at the point* $\mathfrak{p}$ as the group

$$H_q(R, \mathfrak{p}) = H_q(\Theta) = H_q(\mathsf{H}/\mathsf{P}).$$

The following theorem, which is especially useful in applications, allows the local homology groups $H_q(R, \mathfrak{p})$ to be determined by means of an arbitrary simplicial decomposition K of R, for which $\mathfrak{p}$ is not necessarily a vertex.

24.1 Theorem: *The local homology groups of a polyhedron* $R = |\mathsf{K}|$ *at a point* $\mathfrak{p}$ *with the carrier simplex* σ *are isomorphic with the homology groups of the star* $\Theta = \mathrm{St}\ \sigma$.

Proof: Suppose the point p has the carrier simplex σ. Then $\mathfrak{p}$ is an interior point of σ. Let $\Theta = \mathrm{St}\ \sigma$ and $\mathsf{H} = \mathrm{Hu}\ \Theta$ and $\mathsf{P} = \mathrm{Bd}\ \Theta$. We now form a subdivision K' of K for which $\mathfrak{p}$ is a vertex, but $\mathsf{K} - \Theta$ remains unchanged. To this end, we project P from $\mathfrak{p}$, expressed more specifically: For every simplex $\mathfrak{p}_0 \ldots \mathfrak{p}_r$ of Θ, we form the simplex $\mathfrak{p}\mathfrak{p}_0 \ldots \mathfrak{p}_r$ and we replace the simplexes of Θ by the simplexes $\mathfrak{p}\mathfrak{p}_0 \ldots \mathfrak{p}_r$ including $\mathfrak{p}$ itself. (See the adjacent Fig. 9 in which the introduced simplexes are indicated by dotted lines.) We thus obtain a subdivision K' of K. If we set $\Theta' = \mathrm{St}_{\mathsf{K}'}\mathfrak{p}$ and $\mathsf{H}' = \mathrm{Hu}_{\mathsf{K}'}\ \Theta'$ and $\mathsf{P}' = \mathrm{Bd}_{\mathsf{K}'}\ \Theta'$, then H' is a subdivision of H, whereas $\mathsf{P}' = \mathsf{P}$. We apply the theorem on the invariance of the relative homology groups to H and H', which in this case asserts that

$$H_q(R, \mathfrak{p}) = H_q(\mathsf{H}'/\mathsf{P}') = H_q(\mathsf{H}/\mathsf{P}).$$

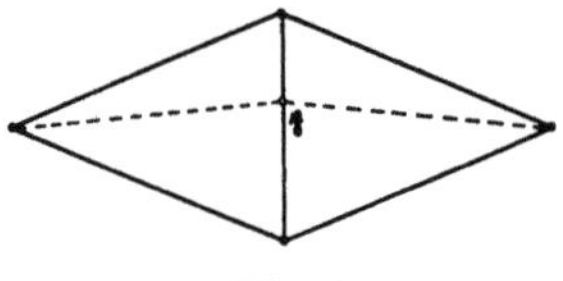

Fig. 9

The last-mentioned group is exactly $H_q(\Theta)$. This completes the proof of the theorem.

Examples of local homology groups: (a) Let T^n be the complex consisting of an n-simplex σ^n with all its faces, and let $T = |\mathsf{T}^n|$. For an interior point $\mathfrak{p}$ of σ^n, according to the last theorem, we have that $H_n(T, \mathfrak{p}) \approx \mathsf{Z}$ and $H_q(T, \mathfrak{p}) \approx 0$ for $q \neq n$. The local homology groups for this $\mathfrak{p}$ are therefore the homology groups of the n-dimensional cells.—(b) Let $R = |\mathsf{K}|$ be an n-dimensional polyhedron, $n > 0$. Then by (a), there exists in R points $\mathfrak{p}$ with $H_n(R, \mathfrak{p}) \approx \mathsf{Z}$, whereas for all points $\mathfrak{p}$ of R for dimensions $q > n$ we have that $H_q(R, \mathfrak{p}) \approx 0$. Together these equations characterize the dimension n of R uniquely. We have thereby obtained a new proof based on homology theory for the topological invariance of the dimension of a polyhedron which supplements our earlier invariant characterization by means of the paving theorem (Volume I, Chapter 10).—We note the results:

24.2 Theorem: *The local homology groups at an interior point of an n-simplex are isomorphic to the homology groups of the n-dimensional cell, namely* $H_n \approx \mathsf{Z}$, $H_q \approx 0$ *for* $q \neq n$.

We now come to the proof of the assertions (1) and (2) posed initially. Assertion (1) is very easy:

24.3 Theorem: *Let R be a polyhedron and $\mathfrak{p}$ be a point of R. In every neighborhood U of $\mathfrak{p}$, there are closed neighborhoods of $\mathfrak{p}$ which permit a simplicial decomposition into a closed star* $\mathsf{H} = \mathrm{Hu}\Theta$ *with* $\Theta = \mathrm{St}\mathfrak{p}$.

Proof: Let K be a simplicial decomposition of R. Let the carrier of $\mathfrak{p}$ in K be σ. Let K' be the normal subdivision if K with the modification that in place of the center of gravity $\hat{\sigma}$ the point $\mathfrak{p}$ is taken as the new vertex interior to σ, as is clearly possible. Then $\mathfrak{p}$ is a vertex of K'. By using a suitable iterated normal subdivision $\mathsf{K}'^{(\nu)}$ of K', we have that the closed star of $\mathfrak{p}$ in $\mathsf{K}'^{(\nu)}$ is contained entirely in U, which was to be shown.

To prove (2), we consider, besides the star Θ just considered with the midpoint $\mathfrak{p}$ and with $\text{Hu}\Theta = \mathsf{H}$ and $\text{Bd}\Theta = \mathsf{P}$, a second simplicial star Θ' about $\mathfrak{p}$ with $\text{Hu}\Theta' = \mathsf{H}'$ and $\text{Bd}\Theta' = \mathsf{P}'$ for which likewise $|\mathsf{H}'|$ is a closed neighborhood of $\mathfrak{p}$ in R. As we know, H and H' may be situated relative to one another in a very complicated way. We have to show that $H_q(\Theta) = H_q(\Theta')$. Now, according to Theorem 13.5, the homology groups of Θ are completely determined by the homology groups of P, and the corresponding fact is valid for Θ'. It thus suffices to show that P and P' have the same homology groups. This is guaranteed by the following theorem, which, when proven, assures the correctness of our definition of the local homology groups $H_q(R, \mathfrak{p})$.

24.4 Theorem: *Let R be a polyhedron and $\mathfrak{p}$ a point of R. Let Θ and Θ' be two simplicial stars about the midpoint $\mathfrak{p}$ with* $\text{Hu}\Theta = \mathsf{H}$, $\text{Bd}\Theta = \mathsf{P}$ *and* $\text{Hu}\Theta' = \mathsf{H}'$, $\text{Bd}\Theta' = \mathsf{P}$ *such that $|\mathsf{H}|$ and $|\mathsf{H}'|$ are closed neighborhoods of $\mathfrak{p}$ in R. Then the polyhedra $|\mathsf{P}|$ and $|\mathsf{P}'|$ belong to the same homotopy type.*

Proof: By a contraction of the star H with respect to the point set $|\mathsf{H}|$ we mean a similarity transformation under which each point $\mathfrak{x} \in |\mathsf{H}|$ is carried over into a point $\mathfrak{y}$ on the rectilinear segment $\mathfrak{x}\mathfrak{p}$ whose distance $d(\mathfrak{y}, \mathfrak{p})$ is a fixed fractional part of the distance $d(\mathfrak{x}, \mathfrak{p})$. We note that interior to the closed simplex of Θ the "rectilinearity" of segments is well-defined, so that the rectilinearity of the segment $\mathfrak{x}\mathfrak{p}$ is also defined in the sense of the metric of H. Under a contraction of H, the complexes and subcomplexes H, Θ and P are mapped into isomorphic, diminished complexes and subcomplexes H_1, Θ_1, P_1.—We first contract $|\mathsf{H}'|$ so that for the image $|\mathsf{H}_1'|$ we have that $|\mathsf{H}_1'| \subset |\mathsf{H}|$ is valid. In this connection, imagine that $|\mathsf{H}|$ as a neighborhood of $\mathfrak{p}$ contains a spherical neighborhood with positive diameter. In the same sense, one stipulates successively H to H_1, H_1' to H_2' and H_1 to H_2 in such a way that $|\mathsf{H}_2| \subset |\mathsf{H}_2'| \subset |\mathsf{H}_1| \subset |\mathsf{H}_1'| \subset |\mathsf{H}|$. The corresponding symbols Θ_1, Θ_1', P_1, P_1', and so on, are understandable without further explanation. Now look at Fig. 10. It is to be noted that "rectilinearity" in the sense of the metric of H need not coincide at all with "rectilinearity" in the sense of the metric of H'. In the stars denoted by primes there is a concept of rectilinearity other than that in the unprimed stars.—We now "project" $|\mathsf{H}_1'| - |\mathsf{H}_2'|$ rectilinearly from $\mathfrak{p}$ onto $|\mathsf{P}_1'|$. In particular, a mapping $\varphi: |\mathsf{P}_1| \to |\mathsf{P}_1'|$ is thus defined. If $\mathfrak{x}$ is an arbitrary point of $|\mathsf{P}_1|$, then, for $0 \leqq t \leqq 1/2$, let $F(\mathfrak{x}, t)$ denote that point which divides the rectilinear segment (in the sense of the metric of H_1') from $\mathfrak{x}$ to $\varphi(\mathfrak{x})$ in the ratio $2t : (1 - 2t)$. Then $F(\mathfrak{x}, 0) = \mathfrak{x}$, $F(\mathfrak{x}, 1/2) = \varphi(\mathfrak{x})$ and $F(\mathfrak{x}, t)$ moves con-

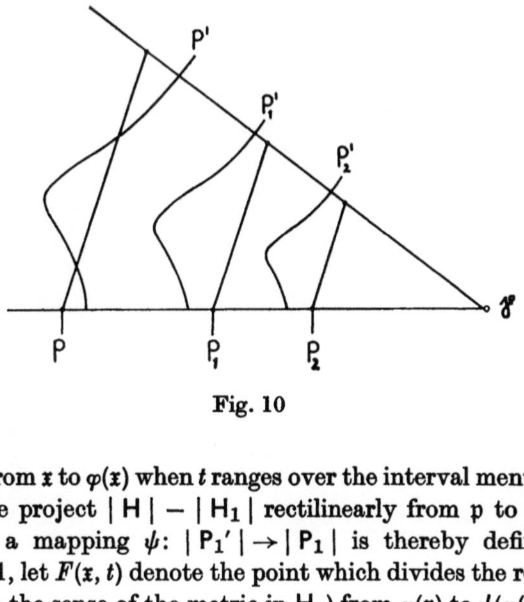

Fig. 10

tinuously from $\mathfrak{x}$ to $\varphi(\mathfrak{x})$ when t ranges over the interval mentioned.—Further, we project $|\mathsf{H}| - |\mathsf{H}_1|$ rectilinearly from $\mathfrak{p}$ to $|\mathsf{P}_1|$. In particular, a mapping ψ: $|\mathsf{P}_1'| \to |\mathsf{P}_1|$ is thereby defined. For $1/2 \leqq t \leqq 1$, let $F(\mathfrak{x}, t)$ denote the point which divides the rectilinear segment (in the sense of the metric in H_1) from $\varphi(\mathfrak{x})$ to $\psi(\varphi(\mathfrak{x}))$ in the ratio $(2t - 1) : (2 - 2t)$. Again $F(\mathfrak{x}, 1/2) = \varphi(\mathfrak{x})$ and we have that $F(\mathfrak{x}, 1) = \psi(\varphi(\mathfrak{x}))$. And $F(\mathfrak{x}, t)$ moves continuously from $\varphi(\mathfrak{x})$ to $\psi(\varphi(\mathfrak{x}))$ when t ranges over the interval mentioned.—We now conceive of $F(\mathfrak{x}, t)$ as a family of homotopic mappings of $|\mathsf{P}_1|$ into $|\mathsf{H}|$

$- |\mathsf{H}_2|$. Let $g\colon |\mathsf{H}| - |\mathsf{H}_2| \to |\mathsf{P}_1|$ be that mapping which projects every point of $|\mathsf{H}| - |\mathsf{H}_2|$ rectilinearly from $\mathfrak{p}$ to $|\mathsf{P}_1|$. Then $g(F(\mathfrak{x}, t))$ is a family of homotopic mappings of $|\mathsf{P}_1|$ into itself. And $g(F(\mathfrak{x}, 0))$ is the identity mapping of $|\mathsf{P}_1|$ onto itself, $g(F(\mathfrak{x}, 1))$ is none other than the composite mapping $\psi\circ\varphi$. Therefore, $\psi\circ\varphi$ is a deformation of the identity mapping of $|\mathsf{P}_1|$. On the basis of symmetry, $\varphi\circ\psi$ is also a deformation of the identity mapping of $|\mathsf{P}_1'|$. Hence, $|\mathsf{P}_1|$ and $|\mathsf{P}_1'|$ belong to the same homotopy type. Since $|\mathsf{P}|$ and $|\mathsf{P}_1|$ and likewise $|\mathsf{P}'|$ and $|\mathsf{P}_1'|$ are homeomorphic, $|\mathsf{P}|$ and $|\mathsf{P}'|$ also belong to the same homotopy type, which was to be proved.

PART IV

DEVELOPMENT OF THE THEORY

Chapter 7

PRODUCTS IN POLYHEDRA

§ 25. The Cohomology Ring

Whereas the homology theory and the cohomology theory of polyhedra up to this point yield one another in dual correspondence, the product theory now to be developed refers exclusively to cohomology groups; there apparently is no corresponding analogy for homology groups. Moreover, other far-reaching differences between homology groups and cohomology groups result in carrying over the theory to more general spaces than polyhedra. This we cannot discuss further in this little volume.

We will assign to each p-dimensional and each q-dimensional cochain c^p and d^q respectively of a simplicial complex K a $(p + q)$-dimensional cochain $c^p \cup d^q = b^{p+q}$ as the product. This product corresponds in many ways to the topological (or cartesian) product. In fact, it can be derived by means of the topological product; however, such a derivation would fall outside the framework of our presentation. We choose a simpler, but somewhat artificial mode of introduction.

25.1 Definition: By an *ordering* $\mathfrak{O}$ *of* K, more specifically an *ordering in the small of the vertices of* K, we mean the following: For each two vertices $\mathfrak{p}_1$ and $\mathfrak{p}_2$ of a simplex K, it is determined which is the earlier and which is the later vertex; if $\mathfrak{p}_1$ and $\mathfrak{p}_2$ belong to two different simplexes of K, then the prescription asserts the same thing in both simplexes. Then K is said to be *ordered* according to $\mathfrak{O}$;

a simplex $\mathfrak{p}_0 \ldots \mathfrak{p}_q$ with the prescribed ordering of vertices is said to be ordered according to $\mathfrak{O}$.

Examples: (1) For the normal subdivision K′ of a complex K, we have already gotten to know a natural ordering of the vertices $\mathfrak{p}'$ in Volume I, § 31, namely the ordering according to the dimension of those simplexes of K whose center of gravity is $\mathfrak{p}'$.—(2) If we enumerate the totality of α_0 vertex points of K in any way from 1 to α_0, then this also yields an ordering in the small in the sense of our definition.

25.2 Definition: Let the simplicial complex K be ordered according to an ordering $\mathfrak{O}$, and let c^p and d^q be two cochains from K. The cochain $b^{p+q} = c^p \cup d^q$, the *cup product* of c^p and d^p, is defined as that $(p + q)$-dimensional cochain which assumes the following value on every ordered simplex $\langle \mathfrak{p}_0 \ldots \mathfrak{p}_{p+q} \rangle$:

$$(\langle \mathfrak{p}_0 \ldots \mathfrak{p}_{p+q} \rangle, c^p \cup d^p) = (\langle \mathfrak{p}_0 \ldots \mathfrak{p}_p, c^p) (\langle \mathfrak{p}_p \ldots \mathfrak{p}_{p+q}, d^q).$$

In order to avoid any misunderstanding, let us add the remark that the product cochain $b^{p+q} = c^p \cup d^q$ is in fact completely determined by this definition although it is given formally only on the $(p + q)$-dimensional simplexes ordered according to $\mathfrak{O}$. Namely, if $s_{p+q} = \langle \mathfrak{p}_{i_0} \ldots \mathfrak{p}_{i_{p+q}} \rangle$ is an arbitrary elementary $(p + q)$-chain, then, for (s_{p+q}, b^{p+q}), we find, according to the general rules for cochains (§ 10, subsection 1), that

$$\begin{aligned}(s_{p+q}, b^{p+q}) &= (\langle \mathfrak{p}_{i_0} \ldots \mathfrak{p}_{i_{p+q}} \rangle, b^{p+q}) \\ &= \text{signum}\,(i_0, \ldots, i_{p+q}) (\langle \mathfrak{p}_0 \ldots \mathfrak{p}_{p+q} \rangle, b^{p+q}),\end{aligned}$$

and this last expression is given in the definition.—In the defining equation, multiplication of integers is utilized on the right; we note that the vertex $\mathfrak{p}_p$ in a certain sense appears as a combining member in both factors.—For example, in the ordered complex of the adjacent figure (see Fig. 11), let c^1_{ik} denote the 1-dimensional elementary cochain which has the value 1 on the directed edge *ik*, but otherwise has the value zero. We denote the corresponding 2-dimensional cochain by c^2_{ike}. Then we have that $c^1_{12} \cup c^1_{23} = c^2_{123}$, and $c^1_{21} \cup c^1_{23} = -c^2_{123}$, but $c^1_{23} \cup c^1_{12} = 0$. The cup product is accordingly seen to be very unsymmetrical. Its simple properties will show up first in the transition to the cohomology classes; it is naturally in no way commutative.

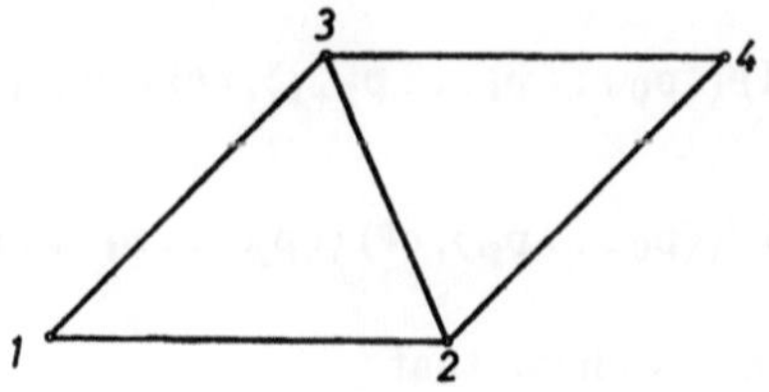

Fig. 11

The cup product satisfies both the distributive laws and the associative law:

$$(c_1^p + c_2^p) \cup d^q = c_1^p \cup d^q + c_2^p \cup d^q,$$

$$c^p \cup (d_1^q + d_2^q) = c^p \cup d_1^q + c^p \cup d_2^q,$$

$$(c^p \cup d^q) \cup e^r = c^p \cup (d^q \cup e^r) = c^p \cup d^q \cup e^r.$$

The two distributive laws follow according to the definition from the corresponding property of the Kronecker product; the associative law follows directly from the definition. The cochain e^0 which is defined by $(\langle \mathfrak{p} \rangle, e^0) = 1$ for all $\mathfrak{p} \in \mathsf{K}$ is obviously the unit element for the cup product:

$$e^0 \cup c^p = c^p, \; c^q \cup e^0 = c^q \text{ for all } c^p, c^q.$$

Namely, if we carry these products out according to Definition 25.2, then the first and the second factor, respectively, are equal to 1.

We now come to the important differentiation formula for the cup product.

25.3 Theorem: $\delta(c^p \cup d^q) = \delta c^p \cup d^q + (-1)^p(c^p \cup \delta d^q)$.

Proof: We investigate the value which the cochain assumes on an elementary $(p + q + 1)$-dimensional chain on the left side:

$$(\langle p_0 \ldots p_{p+q+1}\rangle, \delta(c^p \cup d^q)) = (\partial\langle p_0 \ldots p_{p+q+1}\rangle, c^p \cup d^q)$$

$$= \sum_{i=0}^{p+q+1} (-1)^i(\langle p_0 \ldots \hat{p}_i \ldots p_{p+q+1}\rangle, c^p \cup d^q)$$

$$= \sum_{i=0}^{p} (-1)^i(\langle p_0 \ldots \hat{p}_i \ldots p_{p+1}\rangle, c^p)\,(\langle p_{p+1} \ldots p_{p+q+1}\rangle, d^q)$$

$$+ \sum_{i=p+1}^{p+q+1} (-1)^i(\langle p_0 \ldots p_p\rangle, c^p)\,(\langle p_p \ldots \hat{p}_i \ldots p_{p+q+1}\rangle, d^q).$$

On the other hand, we have that

$$(\langle p_0 \ldots p_{p+q+1}\rangle, \delta c^p \cup d^q)$$

$$= (\langle p_0 \ldots p_{p+1}\rangle, \delta c^p)\,(\langle p_{p+1} \ldots p_{p+q+1}\rangle, d^q)$$

$$= (\partial\langle p_0 \ldots p_{p+1}\rangle, c^p)\,(\langle p_{p+1} \ldots p_{p+q+1}\rangle, d^q)$$

$$= \left[\sum_{i=0}^{p+1} (-1)^i(\langle p_0 \ldots \hat{p}_i \ldots p_{p+1}\rangle, c^p)\right](\langle p_{p+1} \ldots p_{p+q+1}\rangle, d^q)$$

and

$$(\langle p_0 \ldots p_{p+q+1}\rangle, c^p \cup \delta d^q)$$

$$= (\langle p_0 \ldots p_p\rangle, c^p)\,(\langle p_p \ldots p_{p+q+1}\rangle, \delta d^q)$$

$$= (\langle p_0 \ldots p_p\rangle, c^p)\,(\partial\langle p_p \ldots p_{p+q+1}\rangle, d^q)$$

$$= (\langle p_0 \ldots p_p\rangle, c^p)\left[\sum_{i=p}^{p+q+1} (-1)^p(-1)^i(\langle p_p \ldots \hat{p}_i \ldots p_{p+q+1}\rangle, d^q)\right].$$

In the third of these 12 lines of formulas, we now let index i run from 0 to $p + 1$ and in the fourth line we let the index run from p to $p + q + 1$; the summands newly adjoined hereby do not change the equation regarding validity, since they cancel each other mutually. The formula to be proved is now immediate.

An important consequence can be derived from the differentiation formula. First we see that the product of two cocycles $z^p \cup y^q$ is again a cocycle. If we replace further z^p and y^q by homologous cocycles, then

$$(z^p + \delta c^{p-1}) \cup (y^q + \delta d^{q-1}) = z^p \cup y^q + z^q \cup \delta d^{q-1} +$$

$$\delta c^{p-1} \cup y^q + \delta c^{p-1} \cup \delta d^{q-1} \simeq z^p \cup y^q,$$

inasmuch as the last three summands are homologous to zero, since, for example, $\delta(z^p \cup d^{q-1}) = \delta z^p \cup d^{q-1} + (-1)^p (z^p \cup \delta d^{q-1})$; therefore, $z^p \cup \delta d^{q-1} = (-1)^p \delta(z^p \cup d^{q-1}) \simeq 0$. The product of a cohomology class from $H^p(\mathsf{K})$ and a cohomology class from $H^q(\mathsf{K})$ yields accordingly, formed representative-wise, a cohomology class of $H^{p+q}(\mathsf{K})$. This product formation among the cohomology classes of different dimensions we denote likewise by the cup product. The cohomology classes of all dimensions form a so-called *graduated ring:* Classes of the same dimension can be added and subtracted as elements of an abelian group, classes of arbitrary dimension can be multiplied one with another, whereby the dimensions add up. This graduated ring is called the *cohomology ring* $H^*(\mathsf{K})$ *of* K. In order to obtain a ring in the usual sense of the word we could form the direct sum of all $H^q(\mathsf{K})$ with $q = 0, \ldots, n$ and therein define products for arbitrary factors. We shall forego doing this within the framework of this presentation.

We have now to investigate the behavior of the cup product under simplicial mappings φ: $\mathsf{K} \rightarrow \bar{\mathsf{K}}$. Let orderings $\mathfrak{O}$ and $\bar{\mathfrak{O}}$ be given in K and $\bar{\mathsf{K}}$ respectively. Let φ be *order preserving*, i.e., if $\mathfrak{p}_1$ precedes $\mathfrak{p}_2$ in $\mathfrak{O}$, then either $\varphi(\mathfrak{p}_1) = \varphi(\mathfrak{p}_2)$ or $\varphi(\mathfrak{p}_1)$ precedes $\varphi(\mathfrak{p}_2)$ in $\bar{\mathfrak{O}}$. For a given simplicial mapping φ: $\mathsf{K} \rightarrow \bar{\mathsf{K}}$, we can always find orders $\mathfrak{O}$ and $\bar{\mathfrak{O}}$ such that φ is order preserving. For this purpose, we need only to choose $\bar{\mathfrak{O}}$ arbitrarily and $\mathfrak{O}$ in such a way that first come all vertices $\mathfrak{p}$ of K which have as image $\varphi(\mathfrak{p})$ the first vertex of $\bar{\mathsf{K}}$, ordered among themselves in arbitrary sequence, that then come all vertices $\mathfrak{p}$ of K which have as image the second vertex of $\bar{\mathsf{K}}$, again ordered among themselves arbitrarily, and so on.—Therefore, let φ be order preserving. We assert that then the homomorphisms $\varphi^{\bullet}: C^q(\bar{K}) \to C^q(K)$ preserve products; in other words, we assert that for each two cochains $\bar{c}^p \in C^p(\bar{K})$ and $\bar{d}^q \in C^q(\bar{K})$ we have that

$$\varphi^{\bullet}(\bar{c}^p \cup \bar{d}^q) = \varphi^{\bullet}(\bar{c}^p) \cup \varphi^{\bullet}(\bar{d}^q) \tag{25.4}$$

is valid.

For the *proof* it suffices to show that the cochain on the left side and the one on the right side have the same value on an arbitrary $(p + q)$-dimensional elementary chain $\langle \mathfrak{p}_0 \ldots \mathfrak{p}_{p+q} \rangle$ ordered according to $\mathfrak{O}$. The left side yields

$$\begin{aligned}(\langle \mathfrak{p}_0 ,,, \mathfrak{p}_{p+q} \rangle, \varphi^{\bullet}(\bar{c}^p \cup \bar{d}^q)) &= (\varphi_{\bullet} \langle \mathfrak{p}_0 \ldots \mathfrak{p}_{p+q} \rangle, \bar{c}^p \cup \bar{d}^q) \\ &= (\langle \varphi(\mathfrak{p}_0) \ldots \varphi(\mathfrak{p}_{p+q}) \rangle, \bar{c}^p \cup \bar{d}^q),\end{aligned}$$

the last of course only if the vertices $\varphi(\mathfrak{p}_0), \ldots, \varphi(\mathfrak{p}_{p+q})$ are all mutually distinct; otherwise zero results. The right side yields

$$\begin{aligned}&(\langle \mathfrak{p}_0 \ldots \mathfrak{p}_{p+q} \rangle, \varphi^{\bullet}(\bar{c}^p) \cup \varphi^{\bullet}(\bar{d}^q)) \\ &\quad = (\langle \mathfrak{p}_0 \ldots \mathfrak{p}_p \rangle, \varphi^{\bullet}(\bar{c}^p)) (\langle \mathfrak{p}_p \ldots \mathfrak{p}_{p+q} \rangle, \varphi^{\bullet}(\bar{d}^q)) \\ &\quad = (\varphi_{\bullet}\langle \mathfrak{p}_0 \ldots \mathfrak{p}_p \rangle, \bar{c}^p) (\varphi_{\bullet}\langle \mathfrak{p}_p \ldots \mathfrak{p}_{p+q} \rangle, \bar{d}^q) \\ &\quad = (\langle \varphi(\mathfrak{p}_0) \ldots \varphi(\mathfrak{p}_p) \rangle, \bar{c}^p) (\langle \varphi(\mathfrak{p}_p) \ldots \varphi(\mathfrak{p}_{p+q}) \rangle, \bar{d}^q),\end{aligned}$$

the last of course only if there occur in each of the last two Kronecker products only distinct vertices. Now if $\varphi(\mathfrak{p}_0), \ldots, \varphi(\mathfrak{p}_{p+q})$ are all mutually distinct, then the equality of both sides is evident according to the definition of the cup product. If two equal ones among them are present, then the left side is zero. Since these points are ordered according to $\bar{\mathfrak{O}}$, also at least two neighboring ones are equal to one another. Hence at least one of the last two Kronecker products and thus the entire right side is zero. Thus (25.4) is proved.—It follows further that also the induced mapping φ^*: $H^q(\bar{K}) \to H^q(K)$ of the cohomology groups is product-preserving, i.e., to the product class of two cocycles from $H^p(\bar{K})$ and $H^q(\bar{K})$ there corresponds under φ^* the product of the image classes in K.

The cup product among the cochains of K depends essentially on the order $\mathfrak{O}$ used in the definition. We will now show that the cup product for cohomology classes does not depend on $\mathfrak{O}$, that it therefore is a relation which depends solely on K itself. For this purpose, we take an arbitrary order $\mathfrak{O}$ of K and the natural order $\mathfrak{O}'$ of the normal subdivision K′ of K. Let η: K′ → K be that displacement which carries each vertex of K′ over into the highest vertex of its carrier in K. If p′ precedes q′ in $\mathfrak{O}'$, then the carrier simplex σ in K of p′ lies on the boundary of the carrier simplex τ in K of q′. The highest vertex of σ thus precedes the highest vertex of τ or it is equal to the latter. That is, η is order preserving. Hence (25.4) is valid with $\eta^{\bullet}$ in place of $\varphi^{\bullet}$. If we choose the cocycles z^p and y^q arbitrarily in K and for the moment denote the cohomology classes of these cocycles by $[z^p]$ and $[y^q]$ respectively, then furthermore

$$\eta^*([z^p] \cup [y^q]) = \eta^*([z^p]) \cup \eta^*([y^q])$$

is valid. According to our previous discussion (cf. Theorem 21.1), η^* is an isomorphism and moreover it is product preserving, it is a *ring isomorphism*, and as such it possesses an inverse η^{*-1}. It follows that

$$[z^p] \cup [y^q] = \eta^{*-1}(\eta^*[z^p] \cup \eta^*[y^q]).$$

On the right in the parentheses there appears a cup product in K′ which is independent of $\mathfrak{O}$. Whereas η depends essentially on $\mathfrak{O}$, η^* is determined according to Theorem 11.4 solely by K and K′ and, therefore, is independent of $\mathfrak{O}$. We have before us accordingly in the last equation a representation of the cup product which is independent of $\mathfrak{O}$. The cup product in K for cohomology classes is therefore determined solely by K.

In order to investigate the behavior of the cup product under an interchange of both factors, we use the independence of the product from the basic order. Let $\mathfrak{O}$ be any ordering of the simplicial complex K and let $\mathfrak{O}'$ be the reverse ordering. We denote by $\cup'$ the cup product upon taking $\mathfrak{O}'$ as the basic order. If $\langle \mathfrak{p}_0 \ldots \mathfrak{p}_{p+q}\rangle$ is ordered according to $\mathfrak{O}$, then the comparison of $d^q \cup c^q$ and $c^p \cup' d^q$ shows that:

$$(\langle \mathfrak{p}_0 \ldots \mathfrak{p}_{p+q} \rangle, d^q \cup c^p) = (\langle \mathfrak{p}_0 \ldots \mathfrak{p}_q \rangle, d^q)\,(\langle \mathfrak{p}_q \ldots \mathfrak{p}_{p+q} \rangle, c^p),$$

$$(\langle \mathfrak{p}_{p+q} \ldots \mathfrak{p}_0 \rangle, c^p \cup' d^q) = (\langle \mathfrak{p}_{p+q} \ldots \mathfrak{p}_q \rangle, c^p)\,(\langle \mathfrak{p}_q \ldots \mathfrak{p}_0 \rangle, d^q).$$

The two right members differ by the factor -1 to the power $\binom{p+1}{2} + \binom{q+1}{2}$. If, in the first equation on the left we interchange the sequence of the vertices, then there appears as factor -1 to the power $\binom{p+q+1}{2}$. The product of these two factors is -1 to the power

$$\binom{p+1}{2} + \binom{q+1}{2} + \binom{p+q+1}{2} \equiv pq \bmod 2,$$

which we can easily compute or which we can easily verify by distinguishing the different modulo 4 possible cases of p and q. Therefore, in total, there results

$$d^q \cup c^p = (-1)^{pq}(c^p \cup' d^q).$$

If we apply this to cocycles z^p and y^q and their cohomology classes, then according to what has been proved the difference between the two cup products $\cup$ and $\cup'$ vanishes and we obtain the equation

$$z^q \cup y^p \sim (-1)^{pq}(y^p \cup z^q). \tag{25.5}$$

Therefore, according to this equation, the factors of the cup product of cohomology classes commute. We now summarize the preceding results.

25.6 Theorem: *The cohomology classes of all the cohomology groups $H^q(\mathsf{K})$ of a simplicial complex K form by means of the cup product a ring with unit element which is graduated by the dimension q, which is distributive, associative, but in general not commutative. A simplicial mapping φ: $\mathsf{K} \to \bar{\mathsf{K}}$ induces a product-preserving homomorphism φ^*: $H^q(\bar{\mathsf{K}}) \to H^q(\mathsf{K})$.*

In conclusion, we show that the cup product for cohomology

classes is determined solely by the polyhedron $R = |\mathsf{K}|$ and does not depend on the simplicial complex K which is used for the representation of R. To this end, we consider again as in § 21 the totality of simplicial decompositions K_ν, K_μ, . . . of R and the transitive groups systems $H^q(\mathsf{K}_\nu)$, . . . with the isomorphism $\beta(\mu, \nu) = \eta^*(\mu, \lambda) \circ \eta^*(\lambda, \nu): H^q(\mathsf{K}_\nu) \to H^q(\mathsf{K}_\mu)$. These η^* are product-preserving isomorphisms; therefore, the same is also valid for $\beta(\mu, \nu)$. Thus, the $H^q(\mathsf{K}_\nu)$ form a ring isomorphic transitive group system and the product formation carries over in a natural way to the cohomology groups $H^q(R)$ as model groups.—Further, let $f: R \to \bar{R}$ be a continuous mapping. If K and $\bar{\mathsf{K}}$ are simplicial decompositions of R and $\bar{R}$ respectively such that f permits a simplicial approximation $\varphi: \mathsf{K} \to \bar{\mathsf{K}}$, then the homomorphisms $\varphi^*: H^q(\bar{\mathsf{K}}) \to H^q(\mathsf{K})$ are product preserving. From φ^* one can directly deduce the model homomorphism f^*. Also $f^*: H^q(\bar{R}) \to H^q(R)$ is product preserving, i.e., f^* is a *ring homomorphism*. Therefore, if $\bar{Z}^p$ and $\bar{Y}^q$ are p-dimensional and q-dimensional cohomology classes of K respectively, then

$$f^*(\bar{Z}^p \cup \bar{Y}^q) = f^*(\bar{Z}^p) \cup f^*(\bar{Y}^q) \tag{25.7}$$

is valid. We can accordingly establish the following fact.

25.8 Theorem: *The homology groups of a polyhedron R form by means of the cup product a graduated ring. A continuous mapping $f: R \to \bar{R}$ of a polyhedron R into another $\bar{R}$ induces a product-preserving homomorphism $f^*: H^q(\bar{R}) \to H^q(R)$.*

With the aid of the cohomology ring of a space R many questions can be investigated which would be unapproachable solely with the use of homology groups and cohomology groups. Let, for example, R be the 4-dimensional *complex projective plane* which is represented by all homogeneous triples (z_0, z_1, z_2) of complex numbers. Their cohomology groups, which we shall not derive in detail, are $H^0 \approx H^2 \approx H^4 \approx \mathsf{Z}, H^1 \approx H^3 \approx 0$; a generator z^2 of H^2 yields a generator of H^4 in the form $z^2 \cup z^2 = z^4$. Let $\bar{R}$ be the 2-dimensional *complex projective line* contained in R and defined by $z_1 = 0$, which is homeomorphic to the 2-sphere and has the cohomology groups $\bar{H}^0 \approx \bar{H}^2 \approx \mathsf{Z}$, $\bar{H}^1 \approx 0$. We assert that $\bar{R}$ *is not a retract of* R. Namely, otherwise the retraction mapping $r\colon R \to \bar{R}$ would yield a subring $r^*(H^*(\bar{R}))$ of the ring $H^*(R)$ which is isomorphic to the ring $H^*(\bar{R})$ (Theorem 23.5). If $\bar{z}^2$ is a generator of $H^2(\bar{R})$, then we should have $r^*(\bar{z}^2) = kz^2$ with the integer $k \neq 0$. Since $\bar{z}^2 \cup \bar{z}^2 = 0$, it would follow that $r^*(\bar{z}^2 \cup \bar{z}^2) = 0$ and $r^*(\bar{z}^2 \cup \bar{z}^2) = r^*(\bar{z}^2) \cup r^*(\bar{z}^2) = k^2(z^2 \cup z^2) = k^2 z^4 = 0$, therefore $z^4 = 0$, which is not the case.

§ 26. The Cap Product

We have yet to introduce a second sort of product; it is a matter here of a generalization of the Kronecker product.

26.1 Definition: Let the simplicial complex K be ordered according to the ordering $\mathfrak{D}$. Let $s_{p+q} = \langle \mathfrak{p}_0 \ldots \mathfrak{p}_{p+q} \rangle$ be an elementary $(p+q)$-dimensional chain ordered according to $\mathfrak{D}$ and let d^p be an arbitrary cochain. Then the *cap product* $s_{p+q} \cap d^p$ is defined as a q-dimensional chain as follows:

$$\langle \mathfrak{p}_0 \ldots \mathfrak{p}_{p+q} \rangle \cap d^p = (\langle \mathfrak{p}_0 \ldots \mathfrak{p}_p \rangle, d^p) \, \langle \mathfrak{p}_p \ldots \mathfrak{p}_{p+q} \rangle.$$

For arbitrary chains the cap product is defined by linear extension: For integers g_1, g_2 we have that $(g_1 c^1_{p+q} + g_2 c^2_{p+q}) \cap d^p = g_1 c^1_{p+q} \cap d^p + g_2 c^2_{p+q} \cap d^p$.

Thus, by means of the cap product to each $(p+q)$-dimensional chain and to each p-dimensional cochain of K there is assigned a q-dimensional chain. Whereas the first distributive law already appears in the definition, the second distributive law

$$c_{p+q} \cap (g_1 d_1{}^p + g_2 d_2{}^p) = c_{p+q} \cap g_1 d_1{}^p + c_{p+q} \cap g_2 d_2{}^p$$

follows directly from the corresponding property of the Kronecker product. Furthermore, the following theorem is valid.

26.2 Theorem: $(c_{p+q+r} \cap d^p) \cap e^q = c_{p+q+r} \cap (d^p \cup e^q)$.

Proof: The left side of this equation yields for an elementary chain $c_{p+q+r} = \langle \mathfrak{p}_0 \ldots \mathfrak{p}_{p+q+r} \rangle$ in the ordering $\mathfrak{O}$ that

$$(\langle \mathfrak{p}_0 \ldots \mathfrak{p}_{p+q+r} \rangle \cap d^p) \cap e^q =$$

$$(\langle \mathfrak{p}_0 \ldots \mathfrak{p}_p \rangle, d^p) \langle \mathfrak{p}_p \ldots \mathfrak{p}_{p+q+r} \rangle \cap e^q$$

$$= (\langle \mathfrak{p}_0 \ldots \mathfrak{p}_p \rangle, d^p) (\langle \mathfrak{p}_p \ldots \mathfrak{p}_{p+q} \rangle, e^q) \langle \mathfrak{p}_{p+q} \ldots \mathfrak{p}_{p+q+r} \rangle$$

$$= (\langle \mathfrak{p}_0 \ldots \mathfrak{p}_{p+q} \rangle, d^p \cup e^q) \langle \mathfrak{p}_{p+q} \ldots \mathfrak{p}_{p+q+r} \rangle$$

$$= \langle \mathfrak{p}_0 \ldots \mathfrak{p}_{p+q+r} \rangle \cap (d^p \cup e^q).$$

Clearly the Definition 26.1 yields in the case $q = 0$ essentially the Kronecker product (cf. Definition 9.1):

26.3 Theorem: $\mathrm{In}\,(c_p \cap d^p) = (c_p, d^p)$.

26.4 Theorem: $(c_{p+q} \cap d^p, e^q) = (c_{p+q}, d^p \cup e^q)$.

Proof: $(c_{p+q} \cap d^p, e^q) = \mathrm{In}\,((c_{p+q} \cap d^p) \cap e^q = \mathrm{In}\,(c_{p+q} \cap (d^p \cup e^q)) = (c_{p+q}, d^p \cup e^q)$.

Of particular significance is again the differentiation formula.

26.5 Theorem: $(-1)^p \partial(c_{p+q} \cap d^p) = \partial c_{p+q} \cap d^p - c_{p+q} \cap \delta d^p$.

Proof: If $x = x^{q-1}$ is an arbitrary cochain, then, if the left side is denoted by $(-1)^p A$ and the two summands on the right side by B and $-C$, we have that

$$(A, x) = (\partial(e_{p+q} \cap d^p), x) = (c_{p+q} \cap d^p, \delta x) = (c_{p+q}, d^p \cup \delta x),$$

$$(B, x) = (\partial c_{p+q} \cap d^p, x) = (\partial c_{p+q}, d^p \cup x) = (c_{p+q}, \delta(d^p \cup x)),$$

$$(C, x) = (c_{p+q} \cap \delta d^p, x) = (c_{p+q}, \delta d^p \cup x).$$

The last expression of the middle row is equal to

$$(c_{p+q}, \delta d^p \cup x) + (-1)^p (c_{p+q}, d^p \cup \delta x)$$

according to the differentiation formula for the cap product. It therefore follows that $(B, x) = (C, x) + (-1)^p (A, x)$, $(-1)^p (A, x) = (B - C, x)$. Since this is valid for arbitrary cochains $x = x^{y-1}$, it follows that $(-1)^p A = B - C$, and this is the assertion.

From this differentiation formula entirely analogous inferences can be drawn, as we have inferred them in the preceding section for the cup product: The cap product of a cycle with a cocycle is a cycle, the cap product of a boundary with a cocycle and likewise the product of a cycle with a coboundary is a boundary. Hence, the cap product for chains and cochains induces a cap product for homology classes and cohomology classes. The cap product associates to each pair (a, b), where $a \in H_{p+q}(\mathsf{K})$ and $b \in H^p(\mathsf{K})$, an element of $H_q(\mathsf{K})$.

We have now to investigate the behavior of the cap product under simplicial mappings $\varphi\colon \mathsf{K} \to \bar{\mathsf{K}}$. Again let orderings $\mathfrak{O}$ and $\bar{\mathfrak{O}}$ be given in K and $\bar{\mathsf{K}}$ respectively, and let φ be order-preserving. We assert that for every chain $c_{p+q} \in C_{p+q}(\mathsf{K})$ and every cochain $\bar{d}^p \in C^p(\bar{\mathsf{K}})$ the following relation holds:

$$\varphi_\bullet(c_{p+q} \frown \varphi^\bullet \bar{d}^p) = \varphi_\bullet c_{p+q} \frown \bar{d}^p. \tag{26.6}$$

For the *proof* it suffices to take in place of c_{p+q} an arbitrary $(p+q)$-dimensional chain $\langle \mathfrak{p}_0 \ldots \mathfrak{p}_{p+q} \rangle$ ordered according to $\mathfrak{O}$. The right side thus yields

$$\varphi_\bullet\langle \mathfrak{p}_0 \ldots \mathfrak{p}_{p+q}\rangle \frown \bar{d}^p = \langle \varphi(\mathfrak{p}_0) \ldots \varphi(\mathfrak{p}_{p+q})\rangle \frown \bar{d}_p,$$

but of course only if all the vertices $\varphi(\mathfrak{p}_0), \ldots, \varphi(\mathfrak{p}_{p+q})$ are mutually distinct; otherwise zero results. The left side yields

$$\begin{aligned} &\varphi_\bullet(\langle \mathfrak{p}_0 \ldots \mathfrak{p}_{p+q}\rangle \frown \varphi^\bullet \bar{d}^p) \\ &= \varphi_\bullet[(\langle \mathfrak{p}_0 \ldots \mathfrak{p}_p\rangle, \varphi^\bullet \bar{d}^p)\, \langle \mathfrak{p}_p \ldots \mathfrak{p}_{p+q}\rangle] \\ &= (\varphi_\bullet\langle \mathfrak{p}_0 \ldots \mathfrak{p}_p\rangle, \bar{d}^p)\, \varphi_\bullet\langle \mathfrak{p}_p \ldots \mathfrak{p}_{p+q}\rangle \\ &= (\langle \varphi(\mathfrak{p}_0) \ldots \varphi(\mathfrak{p}_p)\rangle, \bar{d}^p)\, \langle \varphi(\mathfrak{p}_p) \ldots \varphi(\mathfrak{p}_{p+q})\rangle, \end{aligned}$$

again only if all the $\varphi(\mathfrak{p}_0), \ldots, \varphi(\mathfrak{p}_{p+q})$ are mutually distinct. In this case the equality of the right and left sides is evident. Otherwise we infer exactly as in the proof of (25.4) that at least one of the last two factors is zero; then both sides are zero. This completes the proof of (26.6).—A corresponding equation is valid if we replace c_{p+q} and $\bar{d}^p$ by the homology classes from $H_{p+q}(\mathsf{K})$ and $H^p(\bar{\mathsf{K}})$ respectively and instead of $\varphi_\bullet$ and $\varphi^\bullet$ the induced homomorphisms φ_* and φ^* respectively.

The cap product among chain and cochains of K depends in an essential manner on the ordering $\mathfrak{O}$ used in the definition. We will now show that the cap product for homology classes and cohomology classes does not depend on $\mathfrak{O}$, that it is a relation which is determined solely by K itself. To this end, again let $\mathfrak{O}$ be an arbitrary ordering of K and let $\mathfrak{O}'$ be the natural ordering of the normal subdivision K', further let η be that displacement of K' to K which assigns to each vertex of K' the highest vertex of its carrier simplex. Now, as we know from the preceding section, η is order-preserving; therefore (26.6) is valid with $\eta_\bullet$ and $\eta^\bullet$ in place of $\varphi_\bullet$ and $\varphi^\bullet$ respectively. Furthermore, if we choose the cycle $z'_{p+q} \in C_{p+q}(\mathsf{K}')$ and the cocycle $y^p \in C^p(\mathsf{K})$ arbitrarily and for the moment denote the corresponding homology class of $H_{p+q}(\mathsf{K}')$ and the corresponding cohomology class of $H^p(\mathsf{K})$ by $[z'_{p+q}]$ and $[y^p]$ respectively, then

$$\eta_*([z'_{p+q}] \cap \eta^*[y^p]) = \eta_*[z'_{p+q}] \cap [y^p]$$

is valid. Since η_* and η^* are isomorphisms, the inverses η_*^{-1} and η^{*-1} exist. We set $\eta_*[z'_{p+q}] = [z_{p+q}]$, $[z'_{p+q}] = \eta_*^{-1}[z_{p+q}]$ with suitable $z_{p+q} \in C_{p+q}(\mathsf{K})$. Then upon interchanging the sides,

$$[z_{p+q}] \cap [y^p] = \eta_*(\eta_*^{-1}[z_{p+q}] \cap \eta^*[y^p])$$

holds. Here, on the right stands the cap product in K', which is independent of $\mathfrak{O}$. And η_* and η^* are, in contrast to $\eta_\bullet$ and $\eta^\bullet$, according to Theorem 11.4, likewise independent of $\mathfrak{O}$. Therefore, we have a representation of the cap product before us which is independent of $\mathfrak{O}$. Therefore, the cap product in K for homology classes and cohomology classes is in fact determined solely by K.

Finally we show that also the cap product for homology classes and cohomology classes is established solely by the polyhedron $R = | \mathsf{K} |$, but does not depend on the simplicial complex K used for the representation of R. The proof proceeds similarly to the corresponding proof for the cup product in the preceding section. If K_ν and K_μ are two simplicial decompositions of $R = | \mathsf{K}_\nu | = | \mathsf{K}_\mu |$, then the isomorphisms $\alpha(\nu, \mu)$ and $\beta(\mu, \nu)$, which according to § 21 join together the homology groups and cohomology groups of K_ν and K_μ, are composed of certain displacements $\eta: \mathsf{K} \to \bar{\mathsf{K}}$ and of the corresponding isomorphisms η_* and η^*. According to 26.6, if z_{p+q} is a cycle from $C_{p+q}(\mathsf{K})$, y^p a cocycle from $C^p(\mathsf{K})$, and $[z_{p+q}]$ and $[y^p]$ denote the corresponding classes, then

$$\eta_*([z_{p+q}] \cap [y^p]) = \eta_*[z_{p+q}] \cap \eta^{*-1}[y^p].$$

In this sense the cap product is preserved by η_* and η^* and correspondingly under $\alpha(\nu, \mu)$ and $\beta(\mu, \nu)$. In this connection, we note the appearance of the reciprocal isomorphism η^{*-1}, which is conditioned by the mutual interaction of homology classes and cohomology classes under the cap product. We can accordingly state the following theorem.

26.7 Theorem: *The homology groups and the cohomology groups of a polyhedron R are related by the cap product, and under appropriate interpretations Theorems* 26.2 *through* 26.4 *hold. Under a continuous mapping $f: R \to \bar{R}$ of R into a polyhedron $\bar{R}$ there holds for every homology class Z_{p+q} of R and every cohomology class $\bar{Y}^p$ of $\bar{R}$ the relation*

$$f_*(Z_{p+q} \cap f^*(\bar{Y}^p)) = f_*(Z_{p+q}) \cap \bar{Y}^p.$$

The last equation follows directly from what has been said from (26.6).

Chapter 8

MANIFOLDS

§ 27. Definitions

Of special interest are those spaces which are locally (i.e., in a neighborhood of each point) like Euclidean R^3 or more generally like R^n with arbitrary natural number n, or more specifically, such spaces for which every point possesses a neighborhood which is homeomorphic to an open n-dimensional Euclidean solid sphere. Such spaces are called *n-dimensional manifolds* M^n. For our purposes, however, the precise definition must be prepared somewhat further. We shall first assume that M^n is a polyhedron, therefore that it possesses a simplicial decomposition. This assumption lies throughout within the framework of our development of algebraic topology, which occupies itself only with polyhedra. But it would in any case be necessary, inasmuch as our knowledge at this time about non-polyhedral manifolds is extraordinarily meager. Accordingly, an n-dimensional manifold M^n should be defined as a polyhedron in which every point has a neighborhood which is homeomorphic to an open n-dimensional solid sphere. This definition also is not yet entirely satisfying insofar as we do not possess at this time a general method for distinguishing for a given simplicial complex whether the neighborhood of a point, for instance the open star of a vertex, is homeomorphic or not to the n-dimensional solid sphere. Also the most extensively developed methods of contemporary topology indicate for this no practically accessible path. On the other hand, for the theory of manifolds to be developed in the sequel, we do not need the complete homeomorphism of suitable neighborhoods of a point $\mathfrak{p}$ with the n-dimensional solid sphere. Actually, we need to know only that the local homology groups at every point $\mathfrak{p}$ of a space are exactly those of the n-dimensional solid sphere. The local homology groups of a polyhedron are, according to § 24, computable directly as the homology groups of the stars St σ of its simplexes σ. Hence we take as basis the following definition for our further considerations:

27.1 Definition: A polyhedron $M = M^n$ is called an *n-dimensional homology manifold* if its local homology groups $H_q(M^n, \mathfrak{p})$ coincide at each of its points $\mathfrak{p}$ with those of the n-cells (i.e., with those of the n-dimensional solid sphere) in an interior point, hence, if $H_n(M^n, \mathfrak{p}) \approx \mathsf{Z}$ and $H_q(M^n, \mathfrak{p}) \approx 0$ hold for $q \neq n$.

According to § 24, a polyhedron, every point of which possesses a neighborhood which is homeomorphic to the open n-dimensional solid sphere, has at every point the homology groups required in the definition. It is thus certainly a homology manifold in the sense of the definition. There exist still other homology manifolds which however are of no special interest for us—since we shall always hold onto this definition in the sequel, we speak henceforth briefly of manifolds, not in detail of homology manifolds.—According to Theorem 24.1, the local homology groups of a polyhedron are the homology groups of the stars St σ of its simplexes σ. The stars St σ of an n-dimensional manifold are accordingly (Theorem 24.2) subcomplexes, with the homology groups of an n-cell, i.e., they are themselves n-cells. This makes it possible for us to conceive of the definition of a manifold purely combinatorially, as is carried out in the following theorem, which is the starting point for our further investigations.

27.2 Theorem: *A polyhedron $M = M^n$ with the simplicial decomposition $\mathsf{M} = \mathsf{M}^n$ is an n-dimensional manifold if and only if the star* St σ *of each simplex σ is an n-dimensional cell, i.e., if H_n* (St σ) $\approx$ Z *and H_q* (St σ) ≈ 0 *for $q \neq n$.*

Examples of manifolds:

(1) A connected 1-dimensional manifold is nothing else than a circumference, which we infer directly from the definition.

(2) We obtain 2-dimensional manifolds in the 2-sphere, in the torus surface, and more generally in the surfaces R_g which were brought in as examples at the end of § 19. That every connected, orientable (see § 9, No. 4) 2-dimensional manifold is homeomorphic with one of the surfaces R_g is the half of the so-called fundamental theorem of surface theory which we shall now formulate and adjoin here. It is of particular interest because it completely solves the homeomorphism problem for 2-dimensional manifolds. Unfortunately, we cannot carry out the not very difficult but tedious proof of this theorem for reasons of space in this little volume. For this reason we point out the presentation in books [7] and [9] of the Literature List. We denote for the moment the connected, orientable or non-orientable 2-dimensional manifolds briefly as surfaces; then the fundamental theorem of surface theory reads as follows.

Fundamental Theorem of Surface Theory: (a) *Every orientable surface F is isomorphic to a 2-sphere with $g \geqq 0$ attached handles; every two surfaces R_g of this sort with distinct g are not homeomorphic.*—(b) *Every nonorientable surface F is homeomorphic to a 2-sphere, for which $k \geqq 1$ disjoint circular disks are replaced by adjoined Möbius bands; every two surfaces B_k of this sort with distinct k are not homeomorphic.*

In the case (a), g is called the *genus* of the surface. And $g = 0$ characterizes the 2-sphere, $g = 1$ the torus, $g = 2$ the two-holed torus R_2. For the Betti numbers we have that $\wp_0 = \wp_2 = 1$, $\wp_1 = 2g$ is valid, for the Euler characteristic $\chi = 2 - 2g$.—In the case (b), $k = 1$ characterizes the closed Möbius band B_1 or the projective plane which is homeomorphic to it. And $k = 2$ characterizes the Klein bottle B_2, which results from the surface sketched in the adjacent Fig. 12 if one

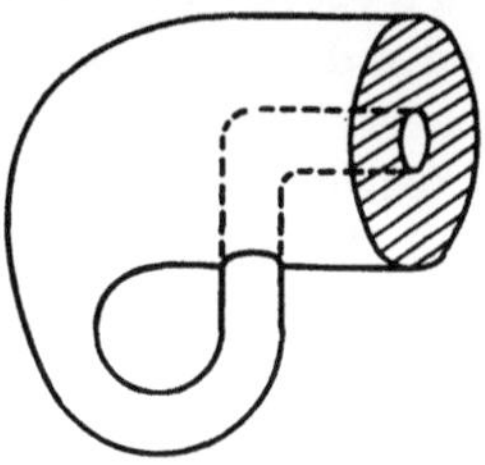

Fig. 12

closes the shaded area by a plane circular ring. It is to be noted that B_2 as well as B_1 is not representable without penetrations into itself in R^3 (but can be in R^4). For the Betti numbers we have that $\wp_0 = 1$, $\wp_2 = 0$, $\wp_1 = k - 1$ is valid, for the Euler characteristic $\chi = 2 - k$. And B_k possesses a 1-dimensional torsion coefficient $e^{(1)} = 2$.

(3) For 3-dimensional manifolds the difficulty pointed out at the beginning of this section still does not exist; in a 3-dimensional triangulated homology manifold the star of every vertex is a 3-dimensional solid sphere as we can recognize on the basis of the fundamental theorem of surface theory without any great difficulty. A complete enumeration of all 3-dimensional manifolds has up to this time not yet been attained. We obtain examples in the 3-sphere and in the topological product of 1- and 2-dimensional manifolds.—We now prove two theorems that a manifold is in any case a pseudomanifold in the sense of Definition 9.7.

27.3 Theorem: *The vertex stars* St $\mathfrak{p}$ *of an n-dimensional manifold* $M^n = |\mathsf{M}|$ *arè (non-closed) orientable n-dimensional pseudomanifolds.*

Proof: That St $\mathfrak{p}$ is a subcomplex of M was already established earlier (§ 3). It remains to show that the stars St $\mathfrak{p}$ are (a) homogeneous n-dimensional, (b) unramified, and (c) strongly connected.—(a) For every simplex $\sigma \in \mathsf{M}$, we have that $H^n(\text{St}\,\sigma) \approx \mathsf{Z}$; therefore, n-dimensional simplexes must occur in St σ. Thus, the dimension of σ is at most n, for otherwise there would occur simplexes of a dimension $>n$ in St σ. Therefore, σ is itself an n-simplex or σ lies on the boundary of one such.—(b) For every $\sigma^{n-1} \in \mathsf{M}$, we have that $H_n(\text{St}\,\sigma^{n-1}) \approx \mathsf{Z}$. From this follows according to Theorem 13.6 the non-ramification of M and hence that of St $\mathfrak{p}$.—(c) We combine all n-simplexes of St $\mathfrak{p}$ with a special n-simplex σ^n of St $\mathfrak{p}$ by means of alternating $(n - 1)$- and n-dimensional incident simplexes of St $\mathfrak{p}$ and all their face simplexes insofar as they lie in St $\mathfrak{p}$ to a homogeneous n-dimensional "substar" Θ_1 of St $\mathfrak{p}$. We assume that St $\mathfrak{p}$ is not strongly connected. Then the remaining n-simplexes of St $\mathfrak{p}$ together with all their faces form, insofar as they lie in St $\mathfrak{p}$, a second non-empty homogeneous n-dimensional substar Θ_2 of St $\mathfrak{p}$, and we have that St $\mathfrak{p} = \Theta_1 \cup \Theta_2$, while $\Theta_1 \cap \Theta_2$ is at most $(n - 2)$-dimensional. Every σ^{n-1} from Θ_1 is incident with exactly two n-simplexes of Θ_1. The same holds for Θ_2. We form in Θ_1 the n-dimensional chain c^1 which is equal to the sum of all n-simplexes of Θ_1 with arbitrary orientation, and analogously the n-dimensional chain c^2 in Θ_2.—A chain is called even if in its representation by elementary chains all the coefficients are even. The chains c^1 and c^2 have the following property: *They have an even boundary* $\partial' c^1$ *and* $\partial' c^2$ *respectively in* St $\mathfrak{p}$ *but they themselves are not even.* Let c be any n-dimensional chain of St $\mathfrak{p}$ with this property. Then $\partial' c = 2d$ and $\partial'\partial' c = 2\partial' d = 0$, $\partial' d = 0$. Since d is $(n-1)$-dimensional, it follows, on account of $H_{n-1}(\text{St } \mathfrak{p}) \approx 0$, that $d \sim 0$, $d = \partial' \bar{c}$ for a suitable chain $\bar{c}$. We have that $\partial'(c - 2\bar{c}) = 0$; therefore, $c - 2\bar{c}$ is a multiple of the basic cycle z^0 of the cell St $\mathfrak{p}$, $c - 2\bar{c} = kz^0$. Since c is not even, we can set $c = z^0 + 2b$ with a certain chain b. Therefore, two chains such as c^1 and c^2 differ by an even chain. But this does not pertain to c^1 and c^2 according to their definition. Hence, our assumption is false, St $\mathfrak{p}$ is strongly connected.—It is thus shown that St $\mathfrak{p}$ is an n-dimensional pseudomanifold. The orientability follows from $H_n(\text{St } \mathfrak{p}) \approx \mathsf{Z}$ according to Theorem 13.4.

27.4 Theorem: *A connected n-dimensional manifold* $M^n = |\mathsf{M}|$ *is a (closed) n-dimensional pseudomanifold.*

Proof: The properties (a) and (b) of a pseudomanifold were already established for M^n in the preceding proof. We prove property (c) again indirectly. Suppose M is not strongly connected. Then M decomposes, as did St $\mathfrak{p}$ in the preceding proof, into two non-empty homogeneous n-dimensional subcomplexes K_1 and K_2 with $\mathsf{K}_1 \cup \mathsf{K}_2 = \mathsf{M}$ and at most $(n-2)$-dimensional intersection $\mathsf{K}_1 \cap \mathsf{K}_2$. Since M is connected, a vertex $\mathfrak{p}$ can be chosen in $\mathsf{K}_1 \cap \mathsf{K}_2$. And St $\mathfrak{p}$ contains n-simplexes from both K_1 and K_2. According to the preceding theorem, they can be joined in the interior of St $\mathfrak{p}$ by means of alternating $(n-1)$-and n-dimensional simplexes. Therefore, the same holds in the interior of M, and this is a contradiction.

On the basis of this theorem, we can speak of orientable and of non-orientable manifolds.

§ 28. Complementary Cell Decompositions

We will show that the barystars (§ 4) of a manifold $M^n = |\mathsf{M}|$ are a cell decomposition of the normal subdivision M' of M. The proof is carried out by an induction inference, which in turn depends on the following theorem. This is valid only for $n \geqq 2$; the case $n = 1$ offers no interest.

28.1 Theorem: *In an n-dimensional manifold $M^n = |\mathsf{M}|$ with $n \geqq 2$ the boundary of a vertex star* St $\mathfrak{p}$ *is a connected $(n-1)$-dimensional manifold.*

Proof: We set St $\mathfrak{p} = \Theta$, Bd $\Theta = \mathsf{P}$. As in Theorem 13.5, we assign to each q-simplex $\sigma^q = \mathfrak{p}_0 \ldots \mathfrak{p}_q \in \mathsf{P}$, the $(q+1)$-simplex $\sigma^{q+1} = \mathfrak{p}\mathfrak{p}_0 \ldots \mathfrak{p}_q \in \Theta'$ $(q = 0, \ldots, n-1)$. This correspondence is one to one and (up to sign, as in Theorem 13.5) incidence-preserving. For a given r-simplex $\sigma^r \in \mathsf{P}$, we have to investigate its star in P, $\mathrm{St}_{\mathsf{P}}\sigma^r$. For $q = 1, \ldots, n$, the correspondence described yields the isomorphisms

$$H_q(\mathrm{St}_{\mathsf{P}}\sigma^r) \approx H_{q+1}(\mathrm{St}_{\Theta}\sigma^{r+1}) \approx H_{q+1}(\mathrm{St}_{\mathsf{M}}\sigma^{r+1}).$$

The last-named homology group is $\approx \mathsf{Z}$ for $q + 1 = n$ and ≈ 0 otherwise. Thus, $H_q(\mathrm{St}_{\mathsf{P}}\sigma^r)$ is determined up to the case $q = 0$. If $r > 0$, then St σ^r has no vertices whatsoever; we have that $H_0(\mathrm{St}_{\mathsf{P}}\sigma^r) \approx 0$. This holds also for $r = 0$, because then $\mathfrak{p}$ is the only vertex of the star, and it is homologous to zero inasmuch as $n \geqq 2$.— Thus P is recognized by virtue of Theorem 27.2 as a manifold. That P is connected was already proved along with the proof of Theorem 27.3.

28.2 Theorem: *The barystars of a manifold $M^n = |\mathsf{M}|$ are cells. More specifically, if σ^q is a q-simplex of* M, *then the corresponding barystar $\tau^p = \tau(\sigma^q)$ is a cell of dimension $p = n - q$.*

Proof: The theorem is trivial for $n = 1$. Suppose it has already been proved for all manifolds of a dimension $< n$. Let the simplex σ^q of the assertion first have a dimension $q > 0$. Let $\mathfrak{p}$ be a vertex of

σ^q and again let $\mathrm{St}_{\mathsf{M}}\mathfrak{p} = \Theta$, $\mathrm{Bd}\,\Theta = \mathsf{P}$. Further, let M' be the normal subdivision of M and $\mathrm{St}_{\mathsf{M}'}\mathfrak{p} = \Theta'$, $\mathrm{Bd}\,\Theta' = \mathsf{P}'$. We assert that the normal subdivision of P is isomorphic to P'. We obtain an isomorphism as follows: If $\sigma^r \in \mathsf{P}$ and $\sigma^{r+1} \in \Theta$ are two simplexes which correspond to one another as in the preceding proof, then we map the vertex $\hat{\sigma}^r \in \mathsf{P}'$ into the vertex $\hat{\sigma}^{r+1}$ of the normal subdivision of Θ; the isomorphism is thereby established.—Let σ^{q-1} be that simplex of P' corresponding to the simplex σ^q. The barystar τ^p belonging to σ^q lies on P'. By virtue of the described mapping, there corresponds to it in the normal subdivision of P the barystar $\bar{\tau}$ from σ^{q-1}. P and hence also the normal subdivision of P is, according to the last theorem, an $(n-1)$-dimensional manifold. By the induction hypothesis, $\bar{\tau}$ is thus a cell of dimension $(n-1)-(q-1) = n-q$. Thus τ^p is also a cell of dimension $p = n - q$. The assertion is thus proved for all cells τ^p which belong to simplexes σ^q with $q > 0$. But for vertex stars τ^n the assertion is already contained in the definition of a manifold, i.e., in Theorem 27.2, applied to M'.

The barystars $\tau^p = \tau(\sigma^q)$ are accordingly cells, and they correspond in a one-to-one fashion to the simplexes σ^q of M. On account of this one-to-one relation and on account of the dimension equation $p + q = n$ the τ^p are called *complementary cells*; more specifically, τ^p is called the *cell complementary to* σ^q. The totality of complementary cells τ^p $(q = 0, \ldots, n; p = n, \ldots, 0)$ form a cell decomposition of M'; it is called the *cell decomposition* $\tilde{\mathsf{M}}$ *complementary to* M. In fact, we have learned earlier (§ 20) that M, and more generally every simplicial complex K, is a cell decomposition of M' and of K' respectively. For manifolds the two cell decompositions M and $\tilde{\mathsf{M}}$ of M' appear in a symmetric, or said better in a complementary, way. From the existence of these two cell decompositions follow the most important theorems on manifolds. However, they can only be explained simply if one orients the complementary cells in an appropriate way, as will be done in the following sections.—In the adjacent figure (see Fig. 13), the cell decompositions M, $\tilde{\mathsf{M}}$ and M' are sketched on a portion of a 2-dimensional manifold: The simplicial decomposition M by the heavy lines, the complementary decomposition $\tilde{M}$ by the dotted lines, and the normal subdivision M′ by all the lines taken together. We search out for each simplex σ^0, σ^1, σ^2 of M the corresponding complementary cell τ^2, τ^1, τ^0.—For notation purposes, we note that one finds in the literature in place

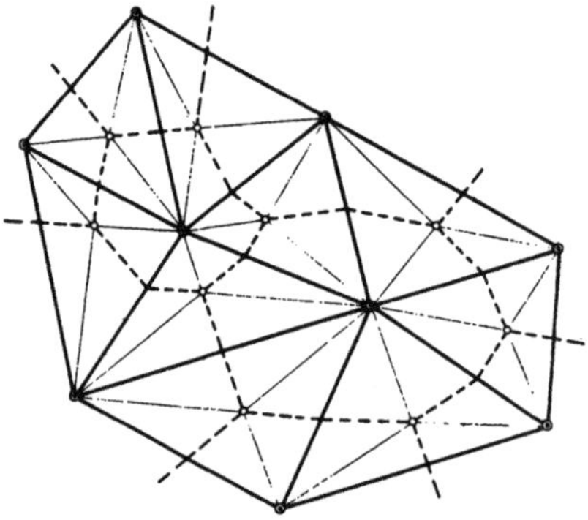

Fig. 13

of the notation "complementary" in general the notation "dual"; we avoid it if possible in order to avoid confusion with the algebraic duality concept of § 15 (dual modules, dual mappings, and so forth).

§ 29. The Poincaré Duality Theorem

Let $M^n = |\,\mathsf{M}\,|$ be an orientable connected n-dimensional manifold with fixed orientation. The assumption of orientability can be avoided in the following investigations if for non-orientable manifolds one will be satisfied with essentially weaker results; but we cannot got into this within the framework of this little volume. As earlier, μ_n and μ_n' denote the basic cycles which are uniquely determined by the orientation of M^n of the simplicial decomposition M and its normal subdivision M′, namely the uniquely determined generators of the n-dimensional homology groups $H_n(\mathsf{M})$ and $H_n(\mathsf{M}')$. The one-to-one correspondence between the simplexes σ^q of M and complementary cells τ^p of $\tilde{M}$ should now be supplemented by means of a correspondence between the oriented simplexes s_q and the oriented complementary cells which are appropriate to be defined. We first give a preliminary sketch which in intuitive form should describe how this correspondence is to be made before we begin the development proper with Theorem 29.1.

Let s_q be an oriented simplex of M, let $s_q' = \alpha\langle \mathfrak{q}_0' \ldots \mathfrak{q}_q' \rangle$ be one of the simplexes of the normal subdivision M' of M, which combine the s_q, and indeed in a natural correspondence of the vertices and with a factor $\alpha = \pm 1$ such that s_q' acquires the orientation determined by s_q, in other words, that s_q' is a summand of ϑs_q with the right sign. A p-dimensional simplex of the cell τ^p complementary to σ^q then has, under the natural ordering of the vertices, the form $\langle \mathfrak{q}_q' \ldots \mathfrak{q}_n' \rangle$. Then $\langle \mathfrak{q}_0' \ldots \mathfrak{q}_q' \ldots \mathfrak{q}_n' \rangle$ is an n-dimensional simplex of M'; if it acquires by means of the factor $\beta = \pm 1$ the orientation of M^n, then we set $d_p' = \alpha\beta\langle \mathfrak{q}_q' \ldots \mathfrak{q}_n' \rangle$ and we denote by d_p the sum of these d_p' if the summation is over all p-dimensional simplexes of τ^p. As we have seen, this d_p is a basic cycle of the cells τ^p.—The cases $q = n$ and $q = 0$ can be described especially easily: To a simplex s_n of M, oriented according to M^n, there belongs the cell d_0 complementary to the center of gravity $\hat{s}_n$ in the positive orientation; to a positively oriented vertex s_0 of M

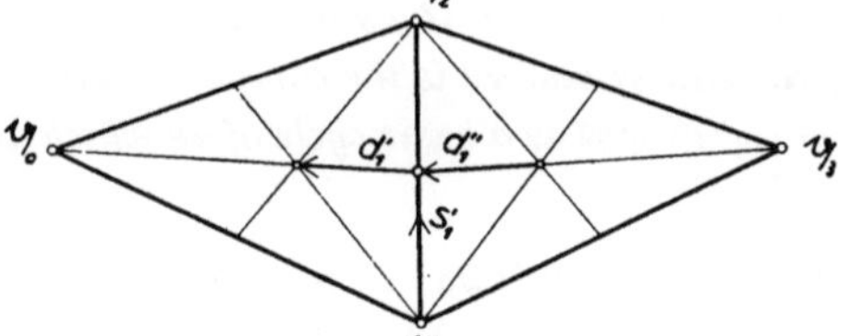

Fig. 14

there belongs the barystar d_n of s_0 in the orientation given by M^n.— The reader should make the construction of d_p clear to himself from the adjacent figure (see Fig. 14), which represents a portion of an

M^2 in that orientation which is counterclockwise. For $q = 1$, say $s_q = s_1 = \langle \mathfrak{q}_1\mathfrak{q}_2 \rangle$, we have that $d_p = d_1 = d_1' + d_1''$.

In order to make the definition of d_p meaningful, we would now have to show (a) that d_p is a cycle and even a basic cycle on τ^p, (b) that d_p is independent of the choice of s_q' in the interior of s_q. It would not be difficult to prove these two assertions directly. But we proceed more rapidly if we make use of the cap product which is closely related to the following investigations, in particular to the Poincaré duality theorem. In this connection, it is expedient to take as basic the elementary cochains s^q instead of the chains s_q and, therefore, to define a correspondence between s^q and d_q, since a better computation is attributed to the complementary behavior of simplex and complementary cell in boundary formation and coboundary formation, which has already been pointed out in Theorem 4.2. The following theorem contains the basic connection between complementary cells d_p and the cap product; it is entirely independent of the preceding preliminary considerations, to which we return at the end of its proof.

29.1 Theorem: *Let $M^n = |\mathsf{M}|$ be an oriented connected n-dimensional manifold. Let $\mathfrak{O}$ be an ordering of M, let $\mathfrak{O}'$ be the natural ordering of the normal subdivision M' of M, let η be the order-preserving displacement of M' to M, which carries every vertex of M' into the last vertex of its carrier simplex. If σ^q is a simplex of M with the corresponding elementary cochain s^q and τ^p is the complementary cell assigned to σ^q, then $\triangle' s^q = \mu_n' \cap \eta^\bullet s^q$ is a basic cycle of τ^p which does not depend on the chosen order $\mathfrak{O}$.*

Proof: Let s^q be given by $(\langle \mathfrak{p}_0 \ldots \mathfrak{p}_q \rangle, s^q) = \alpha = \pm 1$, where the vertices are ordered according to $\mathfrak{O}$. Let $\beta\langle \mathfrak{p}_0' \ldots \mathfrak{p}_n' \rangle$ be a summand of μ_n' (therefore an n-simplex of M') whose vertices are ordered according to $\mathfrak{O}'$; the factor $\beta = \pm 1$ effects the orientation determined by M^n. Then

$$\begin{aligned}\beta\langle \mathfrak{p}_0' \ldots \mathfrak{p}_n' \rangle \cap \eta^\bullet s^q &= \beta(\langle \mathfrak{p}_0' \ldots \mathfrak{p}_q' \rangle, \eta^\bullet s^q)\, \langle \mathfrak{p}_q' \ldots \mathfrak{p}_n' \rangle \\ &= \beta(\eta_\bullet \langle \mathfrak{p}_0' \ldots \mathfrak{p}_q' \rangle, s^q)\, \langle \mathfrak{p}_q' \ldots \mathfrak{p}_n' \rangle.\end{aligned}$$

This expression vanishes if two of the vertices $\eta(\mathfrak{p}_0'), \ldots, \eta(\mathfrak{p}_q')$ are equal. Otherwise we have

$$\beta\langle \mathfrak{p}_0' \ldots \mathfrak{p}_n'\rangle \cap \eta^\bullet s^q = \beta(\langle \eta(\mathfrak{p}_0') \ldots \eta(\mathfrak{p}_q')\rangle, s^q)\, \langle \mathfrak{p}_q' \ldots \mathfrak{p}_n'\rangle.$$

Moreover, since η is order-preserving, we have that $\langle \eta(\mathfrak{p}_0') \ldots \eta(\mathfrak{p}_q')\rangle$ is ordered according to $\mathfrak{O}$. The parentheses is therefore not zero (and in this case equal to α) if and only if $\eta(\mathfrak{p}_0') = \mathfrak{p}_0, \ldots, \eta(\mathfrak{p}_q') = \mathfrak{p}_q$. It follows that $\mathfrak{p}_0' = \mathfrak{p}_0$ and, furthermore, according to the definition of η that $\mathfrak{p}_1'$ is the center of gravity of $\mathfrak{p}_0\mathfrak{p}_1$, that $\mathfrak{p}_2'$ is the center of gravity of $\mathfrak{p}_0\mathfrak{p}_1\mathfrak{p}_2$, and, finally, that $\mathfrak{p}_q'$ is the center of gravity of $\mathfrak{p}_0 \ldots \mathfrak{p}_q$, i.e., that $\mathfrak{p}_q' = \hat{\sigma}^q$. It follows from this that $\triangle' s^q = \mu_n' \cap \eta^\bullet s^q$ has a non-vanishing coefficient for exactly all those p-simplexes which have $\mathfrak{p}_q'$ as first vertex, that is exactly for all p-simplexes of τ^p. These simplexes give to $\triangle' s^q$ the magnitude $\alpha\beta\langle \mathfrak{p}_q' \ldots \mathfrak{p}_n'\rangle$.

We interrupt the proof by three interjectory remarks: (1) $\triangle' s^q$ changes its sign if s^q changes sign or if M^n is re-oriented. In the first case α goes over into $-\alpha$ and β remains unchanged; in the second case α remains unchanged and β goes over into $-\beta$.—(2) The case $q = n$ is especially easy to visualize: Let s^n be oriented as M^n, that is, s^n assumes the value $+1$ if σ^n is oriented according to M^n. In the notation of the proof, $\alpha\langle \mathfrak{p}_0 \ldots \mathfrak{p}_n\rangle$ is then oriented according to M^n. In this case there exists only a single summand of μ_n' which yields a non-vanishing magnitude to $\triangle' s^n$, namely the simplex $\beta\langle \mathfrak{p}_0' \ldots \mathfrak{p}_n'\rangle$, in which $\mathfrak{p}_0' = \mathfrak{p}_0$, $\mathfrak{p}_1'$ is equal to the center of gravity of $\mathfrak{p}_0\mathfrak{p}_1$, and so forth. In addition, $\langle \mathfrak{p}_0' \ldots \mathfrak{p}_n'\rangle$ has the orientation determined by $\langle \mathfrak{p}_0 \ldots \mathfrak{p}_n\rangle$; we must therefore set $\beta = \alpha$. Thus $\triangle' s^n$ is equal to $\langle \mathfrak{p}_n'\rangle$ with the factor $\alpha\beta = \alpha^2 = +1$, i.e., $\triangle' s^n$ is equal to the positively oriented center of gravity $\hat{\sigma}^n$.—(3) Analogously, we recognize in the case $q = 0$: If s^0 has the value $+1$ on the positively oriented vertex $\mathfrak{p}_0$, then $\triangle' s^0$ is the barystar of $\mathfrak{p}_0$, more specifically the sum of the n-simplexes of this barystar in the orientation determined by M^n.

After we have thus assigned to each s^q of M the chain $\triangle' s^q$ from $C_p(\mathsf{M}')$, we now expand this correspondence by means of a linear extension to a mapping $\triangle': C^q(\mathsf{M}) \to C_p(\mathsf{M}')$. Then for every cochain $c^q \in C^q(\mathsf{M})$,

$$\triangle' c^q = \mu_n' \cap \eta^\bullet c^q$$

is valid.

In the boundary formation the operator $\triangle'$ behaves according to Theorem 26.5 as follows:

$$\begin{aligned}\partial \triangle' c^q &= \partial(\mu_n' \cap \eta^\bullet c^q) = (-1)^q[\partial \mu_n' \cap \eta^\bullet c^q - \mu_n' \cap \delta \eta^\bullet c^q] \\ &= (-1)^{q+1}(\mu_n' \cap \eta^\bullet(\delta c^q))\end{aligned}$$

$$\text{(29.2)} \quad \partial \triangle' c^q = (-1)^{q+1} \triangle' \delta c^q,$$

inasmuch as $\partial \mu_n' = 0$. From this equation there will later result decisive consequences; we first apply it to elementary cochains $c^q = s^q$: We have that $\partial \triangle' s^q = (-1)^{q+1} \triangle' \delta s^q$. And δs^q is not zero only on the $(q+1)$-dimensional simplexes which are incident with σ^q. Hence $\triangle'(\delta s^q)$ is not equal to zero only on the boundary $\dot{\tau}^p$. Thus $\partial \triangle' s^q$ is zero on τ^p, i.e., we can, upon application of the boundary formation ∂' to subcomplexes, write $\partial'(\triangle' s^q) = 0$ and $\triangle' s^q$ is a cycle of τ^p. Since all p-simplexes of τ^p appear in it with coefficients ± 1, $\triangle' s^q$ is actually a basic cycle of τ^p.

It remains yet to show that the correspondence $\triangle'$ which is by its definition dependent on $\mathfrak{O}$ remains unchanged if we take as basic instead of $\mathfrak{O}$ another ordering of M. This occurs by induction according to decreasing dimension q. For $q = n$ the assertion is already contained in the interjectory remark (2) of this proof, since in this case $\triangle' s^n$ is the positively oriented point $\hat{\sigma}^n$. Suppose the assertion has already been proved for all dimensions between n and q, q excluded. For an elementary cochain there then holds that $\partial \triangle' s^q = (-1)^{q+1} \triangle' \delta s^q$. Upon changing $\mathfrak{O}$ the right side remains unchanged according to the induction assumption. On the other hand, $\triangle' s^q$ as a basic cycle could change at most its sign, but since $\partial \triangle' s^q \neq 0$ remains unchanged, this is not the case. This completes the proof of Theorem 29.1.

We return briefly to our preparatory considerations at the beginning of this section. As there, let s_q and $s_q' = \langle \mathsf{q}_0' \ldots \mathsf{q}_q' \rangle$ be given.

Then we denote the vertices $\mathfrak{p}_0, \ldots, \mathfrak{p}_q$ of s_q so that $\mathfrak{p}_0 = \mathfrak{q}_0'$, that the center of gravity of $\mathfrak{p}_0\mathfrak{p}_1$ becomes equal to $\mathfrak{q}_1'$, and so forth, and that finally the center of gravity of $\mathfrak{p}_0 \ldots \mathfrak{p}_q$ becomes equal to $\mathfrak{q}_q'$. Then we introduce an ordering $\mathfrak{O}$ into M, in which in all cases $\mathfrak{p}_0$ precedes $\mathfrak{p}_1$, $\mathfrak{p}_1$ precedes $\mathfrak{p}_2$, and so on, finally, $\mathfrak{p}_{q-1}$ precedes $\mathfrak{p}_q$. We consider further the cochain s^q which is given by $(\langle \mathfrak{p}_0 \ldots \mathfrak{p}_q \rangle, s^q) = \alpha$. Then the proof of Theorem 29.1 shows that in $\triangle' s^q$ only the simplexes $\mathfrak{p}_q' \ldots \mathfrak{p}_n'$ of τ^p really appear and that they yield respectively the summands $\alpha\beta\langle \mathfrak{p}_q' \ldots \mathfrak{p}_n' \rangle$. From this it follows that the d_p in the preparatory remark coincide with the $\triangle' s^q$. The assertions (a) and (b) of the preparatory remark are thus proved. We do not need these facts and the relation between the chains s_q and the d_p which are interrelated with the so-called cut-numbers in the sequel.

Theorem 29.1 will now give us the desired orientation of the complementary cells τ^p and, in addition, the chain complex $\mathfrak{C}(\tilde{\mathsf{M}})$ of the complementary cell decomposition $\tilde{\mathsf{M}}$. Now $d_p = \triangle' s^q = \mu_n' \cap \eta^\bullet s^q$ is in any case a basic cycle of τ^p, and we will accordingly make use of d_p for the construction of the oriented cells. For this purpose, we have—according to the general procedure for this in § 18 (compare in particular the subsection before Theorem 18.7)—to assign to each d_p an oriented cell t_p as a new computation element. For each dimension p, we have then to form from these t_p as free generators the groups $C_p(\tilde{\mathsf{M}})$, which then comprize $\mathfrak{C}(\tilde{\mathsf{M}})$. The transition from the t_p so associated to the d_p is given by the subdivision mapping $\mathfrak{H}$; namely, we have that $\mathfrak{H} t_p = d_p$. Then $\mathfrak{H}$ is extended linearly to a mapping $\mathfrak{H}$: $C_q(\tilde{\mathsf{M}}) \to C_q(\mathsf{M}')$. And $\mathfrak{H}$ is a homomorphic and boundary-preserving mapping. We again give the one-to-one relation between the elementary cochains s^q and the complementary cells t_p by means of a new operator $\triangle$: Let

$$t_p = \triangle s^q, \; s^q = \triangle^{-1} t_p.$$

By means of a linear extension on $C^q(\mathsf{M})$ and $C_p(\tilde{\mathsf{M}})$, $\triangle$ and $\triangle^{-1}$ become operators, which map as follows:

$$\triangle : C^q(\mathsf{M}) \to C_p(\tilde{\mathsf{M}}), \; \triangle^{-1} : C_p(\tilde{\mathsf{M}}) \to C^q(\mathsf{M}).$$

And $\triangle$ is one-to-one and maps the basis elements s^q of $C^q(\mathsf{M})$ on the basis elements t_p of $C_p(\tilde{\mathsf{M}})$. And $\triangle$ and $\triangle^{-1}$ are called the *Poincaré operators for chain and cochains* respectively. The intermeshing of the mentioned mappings is illuminated by the adjacent sketch (see Fig. 15). We establish once more that $\tilde{\vartheta} t_p = d_p = \tilde{\vartheta}\triangle s^q = \mu_n' \cap \eta^{\bullet} s^q$ holds, and therefore that our earlier operator $\triangle'$—

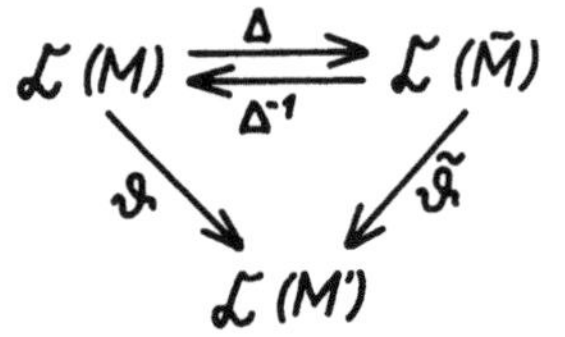

Fig. 15

which no longer will play a role in the sequel as an auxiliary operator —is represented as $\triangle' = \tilde{\vartheta}\triangle$. In general, for every cochain c^q of $C^q(\mathsf{M})$ the relation

$$\tilde{\vartheta}\triangle c^q = \mu_n' \cap \eta^{\bullet} c^q \tag{29.3}$$

is valid.

The bounding relations among the t_p are, according to the definition (cf. § 18), exactly the same as those among the corresponding basic cycles d_p. It therefore follows immediately from Equation 29.2 that for every cochain $c^q \in C^q(\mathsf{M})$

$$\partial\triangle c^q = (-1)^{q+1}\triangle\delta c^q. \tag{29.4}$$

This can also be written briefly as the operation equation $\partial\triangle = (-1)^{q+1}\triangle\delta$. The equation describes the behavior of the operator $\triangle$ with respect to boundary formation and coboundary formation: $\triangle$ carries q-dimensional cochains $c^q \in C^q(\mathsf{M})$ in one-to-one fashion into p-dimensional chains $\triangle c^q = c_p \in C_p(\tilde{\mathsf{M}})$; thereby the coboundary δc^q goes over into the boundary ∂c_p, of course up to the sign $(-1)^{q+1}$. Therefore, $\triangle$ carries the cocycle group $Z^q(\mathsf{M})$ over precisely into the cycle group $Z_p(\tilde{\mathsf{M}})$, likewise the group $B^q(\mathsf{M})$

of the coboundaries into the group $B_q(\tilde{\mathsf{M}})$ of boundaries. The sign $(-1)^{q+1}$ is superfluous here since the mentioned groups contain together with every cycle or cocycle also the opposite cycle or cocycle. Hence $\triangle$ induces, and so does $\triangle^{-1}$, an isomorphism of the homology groups and the cohomology groups as follows:

$$\triangle_*: H^q(\mathsf{M}) \to H_p(\tilde{\mathsf{M}}),\ \triangle^*: H_p(\tilde{\mathsf{M}}) \to H^q(\mathsf{M}).$$

We call $\triangle_*$ and $\triangle^* = \triangle_*^{-1}$ the *Poincaré operators for homology groups and cohomology groups* respectively. They are of great significance for the structure of a manifold. We formulate the result as the next theorem.

29.5 Theorem: (Poincaré Duality Theorem): *In an oriented connected n-dimensional manifold $M^n = |\mathsf{M}|$ the Poincaré duality operator $\triangle$ and the inverse operator $\triangle^{-1}$ induce isomorphisms*

$$\triangle_*: H^q(\mathsf{M}) \to H_q(\tilde{\mathsf{M}});\ \triangle^*: H_p(\tilde{\mathsf{M}}) \to H^q(\mathsf{M}),$$

where $p + q = n$. For the homology groups and cohomology groups of M^n,

$$H^q(M^n) \approx H_q(M^n)$$

holds; $\triangle_*$ maps the left group into the right group isomorphically, and $\triangle^*$ maps the right group into the left group isomorphically.

The proof of the second part of the theorem is immediate: $\tilde{\mathsf{M}}$ possesses, as a cell decomposition of M', the same homology groups and cohomology groups as M' and M. Upon transition from the homology groups and cohomology groups of M and $\tilde{\mathsf{M}}$ to the groups of M^n there result the isomorphisms between $H^q(M^n)$ and $H_p(M^n)$ mentioned in the theorem.—Thus, to each cohomology class $Z^q \in H^q(M^n)$ there corresponds by virtue of $\triangle_*$ a homology class $Y_p = \triangle_* Z^q \in H_p(M^n)$. We call Y_p the *homology class complementary to Z^q*.—For non-orientable manifolds of course there is a corresponding theorem which is perhaps not entirely the best possible; however, we can not go into this for lack of space.

We may designate the Poincaré duality theorem as the most

important foundation of the theory of manifolds. Our further discussion in Chapter 9 is based on this theorem. We first infer several immediate consequences. Let the Betti numbers and the torsion groups of $H_p(M^n)$ be $\wp_p$ and T_p respectively. Then, according to our preceding discussion, the Betti numbers of $H^q(M^n)$ are equal to $\wp_q$ and the cotorsion groups of $H^q(M^n)$ are isomorphic to T_{q-1}. Then the next theorem follows directly from the Poincaré duality theorem.

29.6 Theorem: *In an orientable connected n-dimensional manifold M^n we have, when $p + q = n$, that for the Betti numbers $\wp_p = \wp_q$ is valid; for the torsion groups $T_p \approx T_{q-1}$ is valid and for the cotorsion groups $T^p \approx T^{q+1}$ holds. If n is odd, then $\chi(M^n) = 0$ holds for the Euler characteristic.*

The assertion concerning the Euler characteristic follows by means of the equation $\wp_p = \wp_q$ directly from its definition.—We note that on the basis of this theorem there are at our disposal fewer Betti numbers and torsion numbers for the topological distinction than for arbitrary polyhedra.

A further consequence results from Equation 29.3, if we apply it to a cocycle $c^q = z^q$ and to the corresponding cohomology class Z^q and view it as an equation between cohomology classes and homology classes. Then we have to replace z^q by Z^q; the operators $\mathfrak{H}$ and η do not change the classes, but rather, they reduce to identity operators and can simply be omitted. Moreover, μ_n' and likewise μ_n go over into the generator determined uniquely by means of the oriented manifold M^n of the group $H_n(M^n)$, the basic cycle of M^n, which we shall denote briefly without fear of confusion also by μ_n. There results

$$\triangle_* Z^q = \mu_n \cap Z^q.$$

This is one interpretation of the homology class $\triangle_* Z^q = Z_p$ which is complementary to the cohomology class Z^q: Z_p arises as the cap product of μ^n and Z^q. This result is important insofar as it shows that the Poincaré isomorphism $\triangle_*$ defined on a special simplicial decomposition M by means of the complementary cells of $\tilde{M}$ has in truth an invariant sense independent of M, inasmuch as the cap product has according to § 26 an invariant sense. We establish the following theorem.

29.7 Theorem: ***In an oriented connected n-dimensional manifold M^n the Poincaré duality operator $\triangle_*$ has the representation $\triangle_* Z^q = \mu_n \cap Z^q$ which is independent of the simplicial decomposition of M^n.***

Chapter 9

THE COHOMOLOGY RING OF A MANIFOLD

§ 30. Products in Manifolds

In manifolds the cup product and cap product are interrelated in a simple way, as will now be shown. Again let $M^n = |\mathsf{M}|$ be an oriented connected n-dimensional manifold. Let M' be the normal subdivision, $\tilde{\mathsf{M}}$ the complementary cell decomposition. Let $\vartheta: \mathfrak{C}(\mathsf{M}) \to \mathfrak{C}(\mathsf{M}')$ and $\tilde{\vartheta}: \mathfrak{C}(\tilde{\mathsf{M}}) \to \mathfrak{C}(\mathsf{M}')$ be the corresponding subdivision mappings. Let the displacement $\eta: \mathsf{M}' \to \mathsf{M}$ which appears in the next two theorems be defined by means of a given ordering $\mathfrak{O}$ of M, it carries each vertex of M' into the last vertex of its carrier simplex.

30.1 Theorem: *For every cochain* c^q *of* M,

$$\eta_\bullet \tilde{\vartheta} \triangle c^q = \mu_n \cap c^q$$

is valid.

Proof: According to Equations 29.3 and 26.6 and because $\eta_\bullet \mu_n' = \mu_n$,

$$\eta_\bullet \tilde{\vartheta} \triangle c^q = \eta_\bullet(\mu_n' \cap \eta^\bullet c^q) = \eta_\bullet \mu_n' \cap c^q = \mu_n \cap c^q$$

is valid. In the sequel, the operator $\eta_\bullet \tilde{\vartheta} \triangle: C^q(\mathsf{M}) \to C_p(\mathsf{M})$ plays a role similar to that of the Poincaré operator $\triangle$. It is defined in a somewhat more complicated way than $\triangle$, but, in contrast to $\triangle$, it operates exclusively in M, which for many purposes is more appropriate.

30.2 Theorem: *For arbitrary cochains c^q and d^s of* M,

$$\eta_{\bullet}\mathfrak{H}\triangle c^q \cap d^s = \eta_{\bullet}\mathfrak{H}\triangle(c^q \cup d^s)$$

is valid.

Proof: According to the last theorem and Theorem 26.2, we have that

$$\eta_{\bullet}\mathfrak{H}\triangle c^q \cap d^s = (\mu_n \cap c^q) \cap d^s = \mu_n \cap (c^q \cup d^s) =$$
$$= \eta_{\bullet}\mathfrak{H}\triangle(c^q \cup d^s).$$

We apply this theorem to cocycles $c^q = z^q$ and $d^s = y^s$ and to their cohomology classes Z^q and Y^s and view the equation of the theorem as an equation between homology classes and cohomology classes. Then $\eta_{\bullet}$ and $\mathfrak{H}$ go over into the identity operators and we obtain the next theorem.

30.3 Theorem: *For arbitrary cohomology classes Z^q and Y^s of M^n,*

$$\triangle_* Z^q \cap Y^s = \triangle_*(Z^q \cup Y^s)$$

is valid.

30.4 Theorem: *In a connected oriented n-dimensional manifold the cup product and the cap product for cohomology classes and homology classes can be reduced to one another as follows:*

$$Z_q \cup Y^s = \triangle^*(\triangle_* Z^q \cap Y^s),$$
$$Z_p \cap Y^s = \triangle_*(\triangle^* Z_p \cup Y^s).$$

The first equation results from the preceding theorem by application of $\triangle^*$, the second equation follows if we set $\triangle_* Z^q = Z_p$.—In arbitrary polyhedra there does not exist such a reduction of one product to the other.

For *continuous mappings* of manifolds there result special relations among the homology groups which is in contrast to the case of such mappings of arbitrary polyhedra. Let $M = M^m$ be a connected oriented m-dimensional manifold (with a slight modification of our prior notation); we denote its cohomology classes and homology classes by X^q, Y^q, Z_p, and so forth, where we always have $p + q = m$. Let $N = N^n$ be a connected oriented n-dimensional manifold with the cohomology classes and homology classes U^s, V^s, W_r, and so on, where we always have $r + s = n$. Let $f: M \to N$ be a continuous mapping. The dimension difference $d = m - n$ can be zero, positive or negative. We will place alongside the two mappings induced by f of the homology groups and cohomology groups

$$(\#) \qquad f_*: H_q(M) \to H_q(N), f^*: H^q(N) \to H^q(M)$$

two other mappings which perhaps proceed in the opposite direction. To this end, we form

$$(\#') \qquad \varphi_* = \triangle_* f^* \triangle^*: H_s(N) \to H_{s+d}(M),$$

$$\varphi^* = \triangle^* f_* \triangle_*: H^q(M) \to H^{q-d}(N).$$

We interpret for instance the second equation in $(\#')$: To a cohomology class X^q of M there is first made to correspond, by virtue of $\triangle_*$, the Poincaré operator in M, the complementary homology class $Y_{m-q} = Y_q$, then by virtue of f_* the homology class U_{m-q} in N, finally by virtue of $\triangle^*$, the Poincaré operator in N, the cohomology class $V^{n-(m-q)} = V^{q-d}$ of N. The left side of $(\#)$ is to be viewed in a corresponding way. These mappings φ_* and φ^*, the *reversal mappings* introduced by H. Hopf, therefore proceed in the direction reverse to f_* and f^* respectively; they form for manifolds an important supplement to f_* and f^*, φ_* changes the dimension of the homology groups by $+d$, φ^* changes the dimension of the cohomology groups by $-d$. We restrict ourselves in the following to the mapping φ^*; the corresponding theorems for φ_* are easily formulated. We investigate the behavior of φ^* with respect to product formation.

30.5 Theorem: *For arbitrary cohomology classes Z^q from M and U^s from N,*

$$\varphi^*(Z^q \cup f^* U^s) = (\varphi^* Z^q) \cup U^s$$

holds.

The proof makes use in succession of the Theorems 30.3, 26.7 and 30.4:

$$\Delta^* f_* \Delta_* (Z^q \cup f^* U^s) = \Delta^* f_* (\Delta_* Z^q \cap f^* U^s) = \Delta^* (f_* \Delta_* Z^q \cap U^s)$$
$$= \Delta^* \Delta_* (\Delta^* f_* \Delta_* Z^q \cup U^s) = (\varphi^* Z^q) \cup U^s.$$

In § 9 (immediately after Theorem 9.8) we defined the degree γ of a *simplicial* mapping ψ of an oriented n-dimensional pseudomanifold M^n into another pseudomanifold N^n of the same sort: If μ_n and ν_n are the uniquely determined n-dimensional basic cycles of M^n and N^n, then γ is defined by $\psi_\bullet \mu_n = \gamma \nu_n$.—If we now conceive of μ_n and ν_n as the uniquely determined generators of $H_n(M^n)$ and $H_n(N^n)$, then the *degree γ of a continuous mapping f: $M^n \to N^n$* is defined by $f_* \mu_n = \gamma \nu_n$. The mapping degree γ of a continuous mapping f of two same-dimensional connected oriented n-dimensional manifolds is one of the most important invariants of f.

Regarding this invariant we have the following theorem.

30.6 Theorem: *Under a continuous mapping f of two connected oriented manifolds of the same dimension, f: $M^n \to N^n$, of degree γ, for every cocycle U^s of N^n,*

$$\varphi^* f^* U^s = \gamma U^s$$

is valid.

For the *proof* we draw upon not only the generators μ_n of $H_n(M^n)$ which are uniquely determined by M^n, but also the uniquely determined generators μ^n, μ_0 and μ^0 of $H_n(M^n)$, $H_0(M^n)$ and $H^0(M^n)$ respectively. For the definition of μ^n and μ_0, compare § 10. In a simplicial decomposition M of M^n, μ_n and μ^0 are of course also uniquely representable as a cycle and cocycle respectively, namely μ_n as the sum of all coherent n-simplexes which are oriented according to the orientation of M^n, μ^0 as a cocycle which has the value $+1$ on every vertex of M. For μ^n and μ_0 there does not exist such a unique representation; μ_n can be represented by an arbitrary n-dimensional elementary cochain, μ_0 can be represented by an arbitrary vertex of M.—We obviously have that $\Delta_*\mu^n = \mu_0$, $\Delta^*\mu_0 = \mu^n$. $\Delta^*\mu_n = \mu^0$, $\Delta_*\mu^0 = \mu_n$. Let the corresponding homology classes of N^n be ν_n, ν^n, ν_0, ν^0. Furthermore, according to the definition of γ,

$$\varphi^*\mu^0 = \Delta^*f_*\Delta_*\mu^0 = \Delta^*f_*\mu_n = \Delta^*\gamma\nu_n = \gamma\Delta^*\nu_n = \gamma\nu^0$$

is valid. If we now substitute μ^0 for Z^q in the preceding theorem, then it follows that

$$\varphi^*(\mu^0 \cup f^*U^s) = (\varphi^*\mu^0) \cup U^s,$$

$$\varphi^*f^*U^s = \gamma\nu^0 \cup U^s = \gamma U^s,$$

inasmuch as μ^0 is the identity of the cohomology ring of M^n and ν^0 is the identity of the ring of N^n (Theorem 25.6). This completes the proof of the theorem.

30.7 Theorem: *Let M^n and N^n be connected oriented manifolds. If the degree γ of the continuous mapping $f: M^n \to N^n$ is different from 0, then, for the Betti numbers of all dimensions q,*

$$\wp_q(M^n) \geq \wp_q(N^n)$$

is valid.

Proof: Let U_i^q $(i = 1, \ldots, r)$ be linearly independent elements of $H^q(N^n)$. If the elements $f^*U_i^q = Z_i^q$ of $H^q(M^n)$ were linearly independent, then the elements $\varphi^*Z_i^q = \varphi^*f^*U_i^q = \gamma U_i^q$ would also be linearly independent, which obviously is not the case. This implies the assertion.—According to this, there is, for example, no continuous mapping with a degree $\gamma \neq 0$ of the 2-sphere onto the 2-dimensional torus.

30.8 Theorem: *Let M^n and N^n be connected oriented manifolds. If M^n can be mapped with a degree $\gamma = \pm 1$ into N^n, then the cohomology ring $H^*(M^n)$ contains a subring which is isomorphic to $H^*(N^n)$; if M^n and N^n can be mapped with a degree ± 1 into one another, then the cohomology rings $H^*(M^n)$ and $H^*(N^n)$ are mutually isomorphic.*

Proof: In the first-mentioned case, let $f\colon M^n \to N^n$ have mapping degree $\gamma = \pm 1$. Then for every dimension s there exist homomorphisms f^* and φ^* of the s-dimensional cohomology groups, $\varphi^*\colon H^s(M^n) \to H^s(N^n)$ and $f^*\colon H^s(N^n) \to H^s(M^n)$. For every homology class $U^s \in H^s(N^n)$ we have, according to Theorem 30.6, that $\varphi^* f^* U^s = \pm U^s$ is valid. If the positive sign appears here, then $\varphi^* f^*$ is the identity homomorphism, from which it follows group-theoretically (Theorem 23.3) that f^* is monomorphic and φ^* is epimorphic and that $H^s(M^n)$ has with $H^n(N^n)$ isomorphic direct summands. It then follows from the fact that f^* is product preserving that the product ring $H^*(M^n)$ has a subring which is isomorphic to the product ring $H^*(N^n)$.—However, if the negative sign appears above, then we can infer in the same way; we verify directly that the group-theoretical deduction method of Theorem 23.3 remains applicable if $\varphi^* f^* U^s = -U^s$.—Further, let $\wp_s$ be the Betti number and t_s the order of the cotorsion group of $H^s(M^n)$ and let $\wp_s'$ and t_s' be the corresponding magnitudes for $H^s(N^n)$ or for the direct summand $f^*(H^s(N^n))$ isomorphic with this group of $H^s(M^n)$. On the basis of the fundamental theorem of abelian groups with a finite number of generators (Theorem 16.4), we recognize that $\wp_s \geq \wp_s'$ (which moreover also follows from the preceding theorem) and that $t_s \geq t_s'$ holds, further, that $H^s(M^n) = f^*(H^s(N^n))$ if and only if $\wp_s = \wp_s'$, $t_s = t_s'$ hold simultaneously. Now if M^n and N^n can be mapped with a degree ± 1 into one another, then by reasons of symmetry the inequalities $\wp_s' \geq \wp_s$, $t_s' \geq t_s$ are valid, therefore the equality sign holds both times. Hence $H^s(M^n)$ and $H^s(N^n)$ are isomorphic. This completes the proof of the theorem.

§ 31. Product Matrices

We now investigate the product formation between the elements Z^p and Y^p of the cohomology groups $H^p(M)$ and $H^s(M)$ of a manifold $M = M^n$ somewhat more precisely in the special case $s = q = n - p$. Then $Z^p \cup Y^q$ is of the form $k\mu^n$ with an integer k; μ^n denotes the uniquely determined generator of the cyclic group

$H^n(M)$. We assert that this product depends only on the reduced residue classes of Z^p and Y^q, i.e., only on the classes of Z^p and Y^q in the reduced cohomology groups $\mathscr{H}^p(M)$ and $\mathscr{H}^q(M)$ (cf. § 17). Namely, if U^p and V^q are torsion elements of $H^p(M)$ and $H^q(M)$ respectively, i.e., of finite order within these groups, then we have that

$$(Z^p + U^p) \cup (Y^q + V^q) = Z^p \cup Y^q + Z^p \cup V^q + U^p \cup Y^q + U^p \cup V^q.$$

Since, for example, mV^q vanishes for appropriate integral $m \neq 0$, $m(Z^p \cup V^q) = Z^p \cup mV^q$ also vanishes; therefore, $Z^p \cup V^q$ is itself a torsion element, and the reduced cohomology class of $Z^p \cup V^q$ is the zero class. The same holds for the third and fourth summands on the right in the above equation. Therefore the cup product assigns to each reduced cohomology class of $\mathscr{H}^p(M)$ and a class of $\mathscr{H}^q(M)$ with $p + q = n$ uniquely a product of the form $k\mu^n$. Hence we shall consider exclusively for the remainder of this section in place of the homology groups and cohomology groups $H_q(M)$ and $H^q(M)$ the reduced homology groups and cohomology groups $\mathscr{H}_q(M)$ and $\mathscr{H}^q(M)$. We agree that upper case Latin letters such as Z^p, Y^q from now on always denote reduced residue classes, i.e., elements of $\mathscr{H}^p(M)$ or $\mathscr{H}^q(M)$. We denote the cup product of two reduced classes Z^p and Y^q simply by $Z^p \cup Y^q$.

The product between reduced cohomology classes can be reduced essentially to the Kronecker product, as the following theorem shows. In this connection, we observe that also the Kronecker product depends only on the reduced classes, as we have already shown in Theorem 17.5.

31.1 Theorem: *For reduced cohomology classes Z^p and Y^q with $p + q = n$ of a connected oriented n-dimensional manifold M^n,*

$$Z^p \cup Y^q = (\triangle_* Z^p, Y^q)\mu^n$$

holds.

Proof: According to Theorem 30.4, we first have that $Z_p \cup Y^q =$

$\triangle^*(\triangle_* Z^p \cap Y^q)$. Further, the formula $(c_q, d^q) = \text{In}\ (c_q \cap d^q)$ of Theorem 26.3, applied to cycles c_q and cocycles d^q and their reduced classes $\triangle_* Z^p$ and Y^p of the manifold M^n, asserts that $\triangle_* Z^p \cap Y^q = (\triangle_* Z^p, Y^q)\mu_0$. It then follows that

$$Z^p \cup Y^q = \triangle^*((\triangle_* Z^p, Y^q)\mu_0) = (\triangle_* Z^p, Y^q)\mu^n,$$

which was to be shown.

We now make use of the fact that the reduced groups are free abelian groups which possess linearly independent bases. We choose in each group $\mathscr{H}^q(M)$ a basis which is described as the $\wp_q$-membered column vector $\mathfrak{b}^q$. For $\mathfrak{b}^q$ there exists according to Theorem 17.5 a uniquely determined dual basis $\mathfrak{b}_q$ of $\mathscr{H}_q(M)$ which is characterized by the matrix equation $(\mathfrak{b}_q, \mathfrak{b}^{q\mathsf{T}}) = E$ (unit matrix). Upon application of the Poincaré operator $\triangle^*\colon \mathscr{H}_q(M) \to \mathscr{H}^p(M)$, which does indeed represent an isomorphism, there results from $\mathfrak{b}_q$ a basis $\tilde{\mathfrak{b}}^p$ of $\mathscr{H}^p(M)$. It is called the *basis* $\tilde{\mathfrak{b}}^p$ *of* $\mathscr{H}^p(M)$ *complementary to the basis* $\mathfrak{b}^q$ *of* $\mathscr{H}^q(M)$. According to the preceding theorem, we have, on account of $\triangle^*\mathfrak{b}_q = \tilde{\mathfrak{b}}^p$, $\triangle_*\tilde{\mathfrak{b}}^p = \mathfrak{b}_q$, that

$$\tilde{\mathfrak{b}}^p \cup \mathfrak{b}^{q\mathsf{T}} = \triangle^*\mathfrak{b}_q \cup \mathfrak{b}^{q\mathsf{T}} = (\mathfrak{b}_q, \mathfrak{b}^{q\mathsf{T}})\mu^n = E\mu^n$$

is valid. Moreover, $\tilde{\mathfrak{b}}^p$ is uniquely determined by this property. Namely, if $\mathfrak{c}^p = P\tilde{\mathfrak{b}}^p$ with the unimodular matrix P of $\wp_p = \wp_q$ rows and columns is another basis of $\mathscr{H}^p(M)$ with $(\mathfrak{c}^p, \mathfrak{b}^{q\mathsf{T}}) = E$, then it follows that

$$E = (\mathfrak{c}^p, \mathfrak{b}^{q\mathsf{T}}) = (P\tilde{\mathfrak{b}}^p, \mathfrak{b}^{q\mathsf{T}}) = (P(\tilde{\mathfrak{b}}^p, \mathfrak{b}^{q\mathsf{T}}) = P,$$

and therefore $P = E$. The reader should convince himself of the fact that the removal of the matrix P from the parentheses is correct. Therefore the following theorem is valid.

31.2 Theorem: *For every basis* $\mathfrak{b}^q$ *of the reduced cohomology group* $\mathscr{H}^q(M)$ *there exists a uniquely determined basis* $\tilde{\mathfrak{b}}^p$ *of* $\mathscr{H}^q(M)$ *with the property that*

$$\tilde{\mathfrak{b}}^p \cup \mathfrak{b}^{q\mathsf{T}} = E\mu^n,$$

namely the complementary basis to $\mathfrak{b}^q$.

We have all the products between p- and q-dimensional reduced cohomology classes in hand if we know all the products of elements of a basis $\mathfrak{b}^p$ of $\mathscr{H}^p(M)$ with elements of a basis $\mathfrak{b}^q$ of $\mathscr{H}^q(M)$. We combine these products into the product matrix $\Pi_{p,q}$ which accordingly is defined by the equation

$$\mathfrak{b}^p \cup \mathfrak{b}^{q\mathsf{T}} = \Pi_{p,q}\mu^n.$$

It is a square matrix with $\wp_p = \wp_q$ rows and integer entries. And $\Pi_{p,q}$ has the following properties (1) to (5):

(1) $$\Pi_{q,p} = (-1)^{pq}\Pi^{\mathsf{T}}_{p,q}.$$

This follows directly from the commutative law of the cup product (25.5) if we subject the definition equation for $\Pi_{p,q}$ to a transposition; we observe that on the right the reflected matrix must be taken.

(2) If we take in place of the basis vectors $\mathfrak{b}^q$ for all q other basis vectors $\mathfrak{c}^q = P_q\mathfrak{b}^q$ with unimodular transformation matrix P_q, then we obtain

$$\mathfrak{c}^p \cup \mathfrak{c}^{q\mathsf{T}} = P_p(\mathfrak{b}^p \cup \mathfrak{b}^{q\mathsf{T}})P_q{}^{\mathsf{T}} = P_p\Pi_{p,q}\, P_q{}^{\mathsf{T}}\mu^n.$$

Therefore $\Pi_{p,q}$ *is transformed into* $P_p\Pi_{p,q}P_q{}^{\mathsf{T}}$.

(3) If $\mathfrak{b}^p = Q_p\tilde{\mathfrak{b}}^p$, where Q_p is a unimodular matrix, the so-called *transition matrix*, then it follows that

$$\mathfrak{b}^p \cup \mathfrak{b}^{q\mathsf{T}} = Q_p(\tilde{\mathfrak{b}}^p \cup \mathfrak{b}^{q\mathsf{T}}) = Q_p\mu^n,$$

and therefore $\Pi_{p,q} = Q_p$. *The product matrices are thus unimodular.*

(4) We set $n = 2m$; for even n, m is the mean dimension of M which becomes important later; for odd n, m is a fraction. Let i range over the dimensions $0 \leqq i < m$. Then i and $n - i$ together range for odd n over all dimensions from 0 to n, for even n over all these dimensions with the exception of the middle dimension m. If we choose for the dimension i the bases $\mathfrak{b}^i$ of the groups $\mathscr{H}^i(M)$ arbitrarily, but for the dimensions $n - i$ as bases of the $\mathscr{H}^{n-i}(M)$ the bases $\tilde{\mathfrak{b}}^{n-i}$ complementary to the $\mathfrak{b}^i$, then we obtain that

$$\Pi_{n-i,i}\mu^n = \tilde{\mathfrak{b}}^{n-1} \cup \mathfrak{b}_i{}^{\mathsf{T}} = E\mu^n,$$

$$\Pi_{i,n-i}\mu^n = \mathfrak{b}^i \cup \tilde{\mathfrak{b}}^{n-i\mathsf{T}} = (-1)^{i(n-i)} (\tilde{\mathfrak{b}}^{n-i} \cup \mathfrak{b}^{i\mathsf{T}})$$
$$= (-1)^{i(n-i)} E\mu^n.$$

All these product matrices are therefore equal to the positively or negatively taken unit matrix E. Hence we can say: *By suitable choice of the cohomology bases we can make all product matrices* $\pm E$ *with perhaps the exception of the middle one* $\Pi_{m,m}$.

(5) Finally, we turn to the middle product matrix $\Pi_{m,m}$, denoted briefly by Π_m. It appears only for even $n = 2m$. We have to distinguish two cases: (a) m is odd, therefore $n \equiv 2 \bmod 4$, and (b) m is even, therefore $n \equiv 0 \bmod 4$.

In the case (a), we have according to (1) that $\Pi_m = (-1)^{m^2} \Pi_m^{\mathsf{T}} = -\Pi_m^{\mathsf{T}}$, Π_m is antisymmetric. Going over to determinants, it follows that $(-1)^{\wp_m} = 1$; hence $\wp_m$ is even. Furthermore, the following theorem is valid.

31.3 Theorem: *For a connected n-dimensional oriented manifold* M^n *with* $n = 2m \equiv 2 \bmod 4$ *the middle Betti number* $\wp_m$ *is even; the product matrix* Π_m *can by means of a suitable choice of the basis of* $\mathscr{H}^m(M^n)$ *be brought into the following normal form*

$$\begin{pmatrix} \begin{array}{cc} \cdot & +1 \\ -1 & \cdot \end{array} & & & \\ & \begin{array}{|cc|} \cdot & +1 \\ -1 & \cdot \end{array} & & \\ & & \ddots & \\ & & & \begin{array}{|cc} \cdot & +1 \\ -1 & \cdot \end{array} \end{pmatrix}.$$

The *proof* of the last assertion proceeds similarly to that of Theorem 16.2, it is of course considerably simpler. Let the elements of $\Pi_m = \Pi$ be π_{ik} $(i, k = 1, \ldots, \wp_m = \wp)$. We have $\pi_{11} = 0$ and in general $\pi_{ii} = 0$, but not all π_{1i} vanish since Π is unimodular. Among the product matrices Π which appear for all possible bases of $H^m(M)$ there exist such for which $\pi_{12} \neq 0$ and indeed such for which $\pi_{12} > 0$; we consider such for which π_{12} is positive and minimal. Then all π_{1i} with $i > 2$ are divisible by π_{12} as one sees by the method of the proof of Theorem 16.2, subsection 2. Hence $\pi_{12} = +1$, since otherwise Π would not be unimodular. By basis changes we can then make all π_{1i} with $i > 2$ vanish, which we see according to the method of proof of Theorem 16.2, subsection 3. On account of the skew symmetry of Π then $\pi_{21} = -1$, $\pi_{j1} = 0$ for $j > 2$. Now, by means of further basis changes, we can also make all π_{2i} with $i > 2$ vanish, whereby also all π_{j2} with $j > 2$ become zero.—Thus the left upper two-rowed block of the normal form is produced and to the right and below the block there stand only zeros. By a repeated application of the procedure described there results the normal form and hence the proof of the theorem.

The two-dimensional manifolds $M^2 = R_g$ (sphere with g handles) which we considered at the end of § 19 form a simple example for this theorem. The 1-dimensional cocycles t^1_{1i} and t^1_{2i}, which correspond to the cells τ^1_{1i} and τ^1_{2i}, satisfy precisely the product relations $t^1_{1i} \cup t^1_{2i} = -t^1_{2i} \cup t^1_{1i} = 1$, whereas all other products vanish.

In the case (b), $n \equiv 0 \bmod 4$, m is even, $\Pi_m{}^\mathsf{T} = \Pi_m$, $\Pi_m = \Pi$ is therefore symmetric with $\wp_m = \wp$ rows and columns. The transformation behavior of Π can be described most simply in the language of quadratic forms. If $\mathfrak{x}^\mathsf{T} = (x_1, \ldots, x_\wp)$ is a row vector with $\wp$ variables, then $F = \mathfrak{x}^\mathsf{T}\Pi\mathfrak{x} = \sum \pi_{ik}x_ix_k$ $(i, k = 1, \ldots, \wp)$ is a quadratic form with integer coefficients in these variables. Upon transition to new variables $\mathfrak{x} = P^\mathsf{T}\mathfrak{y}$ or $\mathfrak{x}^\mathsf{T} = \mathfrak{y}^\mathsf{T}P$ with integer unimodular P we have that $F = \mathfrak{y}^\mathsf{T}(P\Pi P^\mathsf{T})\mathfrak{y}$ becomes the corresponding quadratic form in the variables $\mathfrak{y}$. The new coefficient matrix corresponds exactly to the transformation of Π described above under (2). We can thus state the following theorem.

31.4 Theorem: *For a connected n-dimensional oriented manifold M^n with $n = 2m \equiv 0 \bmod 4$, the symmetric product matrix Π_m is the matrix of a quadratic form in $\wp_m$ variables which is assigned in an invariant manner to the manifold up to unimodular transformations.*

The exhibition of invariants and normal forms for such integer quadratic forms is not a simple arithmetic problem.

BIBLIOGRAPHY

Only the most important works are mentioned which can serve to supplement the material treated in this volume insofar as they do not appear in the bibliography in Volume I (*General Topology*).

[1] Alexandroff, P. S., *Einfachste Grundbegriffe der Topologie*, Berlin, 1932.

[2] Alexandroff, P. S., *Die topologischen Dualitätssätze*, Berlin, 1959 (Translated from the Russian).

[3] Cartan, H., Séminaire 1948/49, *Topologie algébrique*, Paris, 1955.

[4] Hilton, P. J., and S. Wylie, *Homology theory*, Cambridge, 1960.

[5] Hocking, J., and G. Young, *Topology*, Addison-Wesley, Reading, 1961.

[6] Lefschetz, S., *Algebraic topology*, AMS Colloquium Publications, New York, 1942.

[7] Lefschetz, S., *Introduction to topology*, Princeton, 1949.

[8] Schubert, H., *Topologie, Eine Einführung*, Stuttgart, 1964.

[9] Seifert, H., and W. Threlfall, *Lehrbuch der Topologie*, Teubner, Leipzig and Berlin, 1934.

[10] Reidemeister, K., *Einführung in die kombinatorische Topologie*, Braunschweig, 1932.

[11] Patterson, E. M., *Topology*, London, 1956.

[12] Wallace, A. H., *An introduction to algebraic topology*, London, 1957.

[13] EILENBERG, S., and N. STEENROD, *Foundations of algebraic topology*, Princeton, 1952.

[14] ALEXANDROFF, P. S., *Combinatorial topology*, I-III, Rochester, 1956, 1957, 1960 (Translated from the Russian).

[15] REIDEMEISTER, K., *Topologie der Polyeder*, Leipzig, 1938.

[16] BOURGIN, D. G., *Modern algebraic topology*, Macmillan, New York, 1963.

[17] GODEMENT, R., *Topologie algébrique et théorie des faisceaux*, Hermann, Paris, 1958.

[18] HU, S. T., *Homotopy theory*, Academic Press, New York, 1959.

[18′] DE RHAM, G., *Variétés differentiables*, Actualités scientifiques et industrielles, No. 1222, Herman, Paris, 1955.

[18″] HODGE, W. V. D., *The theory and applications of harmonic integrals*, Cambridge University Press, London and New York, 1952.

[19] CAIRNS, S. S., *Introductory topology*, Ronald Press, New York, 1961.

[20] NEWMAN, M. H. A., *Elements of the topology of plane sets of points*, Cambridge, (1939).

[21] SPANIER, E. H., *Algebraic topology*, McGraw-Hill, New York, 1966.

[22] DUGUNDJI, J., *Topology*, Allyn and Bacon, Boston, 1966.

[23] HILTON, P., *Homotopy theory and duality*, Gordon and Breach, 1965.

INDEX

Bei Fragen zur Produktsicherheit wenden Sie sich bitte an:
If you have any questions regarding product safety,
please contact:

Walter de Gruyter GmbH
Genthiner Straße 13
10785 Berlin
productsafety@degruyterbrill.com